T0303556

Tree
Sense

Tree
Sense

Ways of thinking about trees

Edited by
**Susette
Goldsmith**

MASSEY UNIVERSITY PRESS

Contents

Part Two
Greening the Anthropocene

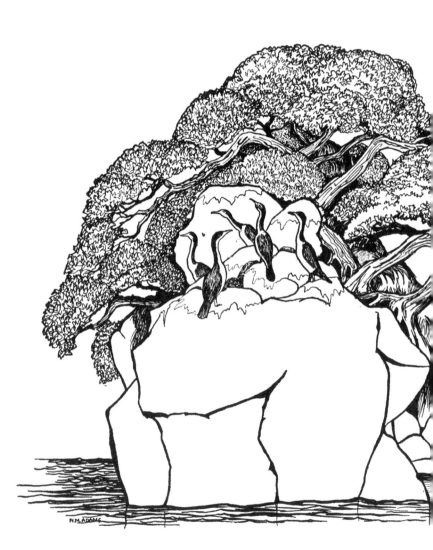

Introduction

___

Susette
Goldsmith

Some time ago now, a curious event occurred right outside my window. My neighbour fired up his chainsaw and proceeded to trim a young, self-sown pōhutukawa that was growing on the edge of the footpath outside his house and leaning towards nearby power lines. It was, he believed, impeding the performance of his computer by brushing against the wires. Another neighbour, passing by, called out to the tree trimmer, urging him to 'cut the whole thing down' as it was spoiling his otherwise unrestricted view.

Alarmed, I leaned out of the window and spoke up in defence of the tree, which, I reasoned, had a right to live and, besides, provided a welcome green frame for my own view. The tree trimmer, somewhat bemused by the interference of his neighbours but keen to satisfy everyone, opted for a compromise — the branches closest to the wires were cut and the others were left alone. The other neighbour had a view with less greenery, I had a view with a ragged frame, the tree trimmer had newly liberated power wires and the tree was still alive.

That might have been an end to it. However, a few weeks later notices from the local council appeared in our letter boxes, reprimanding whoever was responsible for 'illegally pruning' vegetation on public land under Part 17 of the Public Places bylaw. Once more, that might have concluded the affair. But, no. After several more weeks, a council truck arrived at the scene, and several men equipped with a chainsaw and brooms got out. They proceeded to cut down most of what was left of the tree, picked up the branches, stowed them on the truck, swept the footpath around the remaining tree stump and drove away.

In view of the fact that we had *all* been told off for vandalising the tree, and because I was intensely interested in matters arboreal, I emailed the signatory of the infringement notice asking for the reasoning behind the council's action. A month went by before I received a reply from the customer liaison arborist, who apologised for the delay in responding. She explained that it was council policy to remove all trees on public property whose roots threatened the structure of nearby crib walls, and thanked me for my interest in public trees.

A little later, I reported all of this to another of my neighbours, who had been away at the time of the events. She was not at all surprised that the council had cut down the tree, she said, because it was threatening the power lines and, anyway, it wasn't a 'real tree', which I took to mean that it hadn't been planted with human intent. Real tree or not, the pōhutukawa has proved to be resilient. A few years on, the fuss has died down and the stocky little stump has put out many very healthy branches.

The point of this story is not to cast blame:

none of the parties — including the tree — was at fault. My interest in the proceedings stems from the fact that each of the protagonists in our small suburban drama — including the tree — was acting from a vastly different viewpoint. All were, as American environmental historian William Cronon puts it, 'defending their corner of Eden'.[1] The differences in our opinions about the fate of the single pōhutukawa clearly show that there was more than one tree at work here, and that the various trees subscribed to were measured by individual people according to idiosyncratic and anthropocentric values. If, as Cronon argues, our views of nature are important factors in defining who we think we are and the kind of lives we wish to lead, this episode was not just about the tree, but was, in fact, also about us — what we individually believe in and stand for.

That is what this book is about. Although only one tree may be visible in the ground, there may be many invisible trees at work within its drip line, all of which are constructed in the minds of observers according to the meanings and values they hold, and consequently impose, upon the blameless tree. We see trees differently. Some of us affectionately consider them to be sentient beings, while others prioritise their practical attributes of shade and shelter, carbon sequestration, timber production, botanical collection and food. Where some people stand back in awe of the beauty of their autumnal colour changes, others grumble at their leaf-fall. While some champion our indigenous trees, others find superior beauty in exotics, and while some work to protect trees, others labour to fell them.

We may regard trees through any one or any combination of these various lenses, and if this book has a predominant purpose it is to demonstrate to you, the reader, that there are other ways of thinking about trees. Of course, the other way recommended by each of this book's contributors is the manner in which they individually appraise trees. And although we may come from a variety of disciplines and experiences, collectively we are biased; each of us has a deep respect for trees.

Probably, you feel the same way. If you were not interested in the environment and its trees, why else would you select this volume from the bookshop table, library shelf or a friend's desk? To a certain extent this book will preach to the converted. But that's all right, because your thoughts, your opinions and your ways of thinking about trees are valuable. And as we all face up to climate change and the ongoing, alarming challenges to our natural world, we need to stick together, draw strength from one another and preach to the unconverted as well.

E ach of the contributors to this book responded with enthusiasm to my invitation to take part, and I am exceedingly grateful for their commitment to the project and the wonderfully diverse, informative and eloquent chapters they have provided. From the beginning, I envisaged that the structure of the book would be in two parts. The first, 'Needful Dependency', would sing the praises of the various characteristics of trees and remind us of the supportive network that links people and the arboreal world. The second, 'Greening the Anthropocene', would, I hoped, take a lead from the past and provide some guidelines for the future.

Happily, the contributors produced essays and artworks that slipped effortlessly into the two sections. However, this is not a book that requires you to read chapters in the order in which they are presented. Each chapter tackles the topic of trees from a different — and often surprising — viewpoint. Some chapters might attract you more than others, but once you've devoured them I hope you will return for a taste of the rest and — enriched — reflect further on your own relationship with the trees of Aotearoa New Zealand.

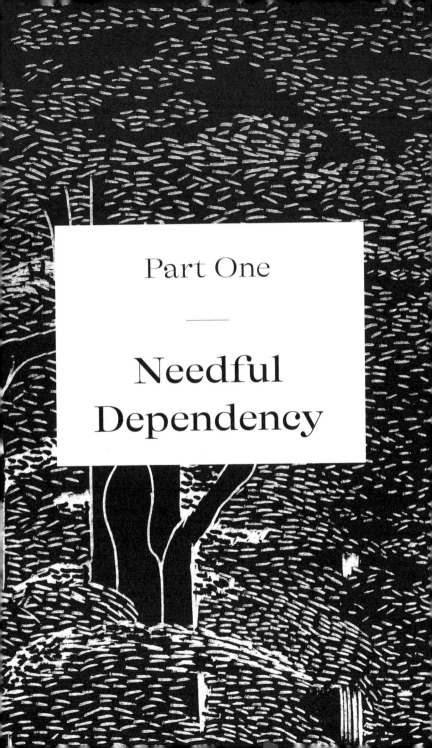

Part One

—

Needful Dependency

Tree breath
and human

———

Elizabeth
Smither

Trees in the garden are expelling
the breath I need. It enters windows
and I breathe out what they breathe in.

How we should love them, never moving
from their posts, withstanding wind,
creaking sometimes, losing limbs

but aiding us with oxygen.
While we climb them to gulp in
the freshest air, the most pure

reason for their being here:
their breath and ours commingling
into needful dependencies.

A Walk
in the
Bush

Philip
Simpson

Let me take you on a walk along the track from Tōtaranui to Goat Bay, in Abel Tasman National Park. As we walk some of us will enjoy the challenge of naming the plants we see, although many will not know any of them. Even so, they are beautiful for their shapes, their colours and their textures. There are the sounds of birds and trickling water and breaking waves. A leaf may fall to the carpet beneath, and we know that this is the cycle of life. This is nature. It is useful to itself and to us, and we can see ourselves as part of it.

We contribute to the latest version of a very long bush history. It evolved without people. We are latecomers, yet we can see the bush as our home, just as other animals do. It wasn't so very long ago that people derived all their requirements from nature. All their food, medicines, building materials, household goods, clothing, minerals, tools and toys came from sources growing, living or located in the immediate surroundings. Despite the beauty inherent in this use, like that in eating a meal from one's own garden, there were ecological repercussions. Fire destroyed forests and hunting caused many species to become extinct almost everywhere in the world, and wherever people migrated they took plants and animals with them, both intentionally and unintentionally. Over time, agriculture and industry replaced our dependence on nature, and trading changed our perspective from local to global.

In the face of this onslaught, nature shrank away. We lost our home. We forgot where things come from. Yet, nearly all the things we started with are still here, although not quite so well off, and our needs are basically just the same. It may well be that some time soon nature will feed culture and we will come to depend on the bush once again. In fact, that is one of the great values of the bush along the Abel Tasman Coast Track: it reminds us of where we came from and it gives us a sense of respect and place. And it is the home of our future.

The species that make up the bush here have been used in many ways over the centuries. I could recall this anywhere in the country, but it is wonderful to walk along a local track — almost in my backyard — and see species after species that have been significant in our history. Knowledge about this history is available in many books and, increasingly, from digital databases. Books I use constantly are Elsdon Best's *Forest Lore of the Maori*,[1] Murdoch Riley's *Maori Healing and Herbal*,[2] and *New Zealand Medicinal Plants*, by the chemists Stanley Brooker and Conrad Cambie and botanist Robert Cooper.[3] Among the many more modern texts are Andrew Crowe's *A Field Guide to the Edible Plants of New Zealand*[4] and Robert Vennell's *The Meaning of Trees*.[5] Very important online sources are *Ngā Tipu Whakaoranga*

— *Māori Plant Use Database*,[6] managed by Manaaki Whenua Landcare Research, and the New Zealand Plant Conservation Network website.[7]

One of the ways in which people use local resources is by naming places according to distinctive features. These names are local identity tags, acting as reminders of cultural history and flags for visitors to recognise. Local people have names for very local features because these identify where particular resources can be found, as well as identifying the rights of access to them. Today, these rights are enshrined under private or public property protected by legal title, but in former times broader whānau or hapū had naming rights over their local rohe. Often a name changes when the land is settled by a new group of people. Obvious examples occur throughout Aotearoa New Zealand, even in the name of the country itself. For example, Whakatū is now Nelson and Mohua is now Golden Bay, although officially the latter is now called Mohua Golden Bay. There is a strong cultural affinity with these names, and when a name is lost the sense of place, of belonging, is diminished.

Tōtaranui is a fine example of the importance of a place name. Tōtara was the most significant tree to most iwi because it was used to make waka and for whakairo. Although widely used by Pākehā as a sympathetic descriptive name for their rural properties, 'Tōtaranui' was seldom used by iwi because places with 'many large tōtara' were numerous throughout the country and therefore not especially distinctive. The name is, however, distinctive here at the start of our walk through the bush. It may have been bestowed by Ngāti Tūmatakokiri, who arrived in the area around 1600, although tangata whenua lived here much earlier.

Ngāti Tūmatakokiri may simply have brought the name with them when they migrated west from Tōtaranui in the Marlborough Sounds. However, it is more likely that the name was given because of the distinctive tōtara population in this new area. Although tōtara would have been scattered throughout the granite country here, the species is actually very rare in Abel Tasman National Park. Granite produces an acidic soil, whereas tōtara prefers fertile alluvium or volcanic ash soil.

Today, there are just a handful of tōtara trees along the coast from Mārahau. The huge storm of December 2012, which precipitated 600 millimetres of rain in 48 hours on the area, caused a massive slip that sheared away a section of the track from Tōtaranui. It also exposed a large lone tōtara on the slip edge; sadly, it was undermined and crashed down to the beach a few months later. However, it was enough to prove that big tōtara were part of the original bush, locally at least. So, the flat where our walk begins is surrounded by country that at one time supported many giant tōtara trees, a locally unusual but extremely valuable resource.

Why were tōtara so important? When the Dutch explorer Abel Janszoon Tasman anchored off Wainui Bay in 1642, he was confronted by a flotilla of more than 20 waka. Each of these was likely carved from a tōtara trunk, although rimu, matai and kahikatea were all sometimes used for this purpose. Tōtara was favoured because the trees are large, and the wood is straight and soft enough to carve relatively easily. But the most important attribute of tōtara was that it is durable in water — the resin inhibits the growth of fungi and bacteria, as well as burrowing sea animals such as *Teredo* 'shipworms'.

Waka were the means of sea travel between settlements, but more especially were used for accessing marine food resources such as fish and other seafood, wetland resources such as tuna and harakeke, and forest resources. They were constantly being built, fixed up and traded. Pākehā who came and settled with the resident iwi immediately learned of the virtues of tōtara. The first houses they built were from axe-split tōtara planks, the roofs sealed with flattened sheets of tōtara bark. The farms were fenced using tōtara posts and battens. The nation shaped its wharves, railways and bridges from tōtara; the surveyors, their pegs. Finally, the trees were virtually gone.

The surviving giants became symbols of conservation, and protest put an end to the logging of native forest. Before that, the bush at Tōtaranui was the epicentre of Pérrine Moncrieff's campaign to establish the national park, which she achieved in 1942. More than any other tree, tōtara carries the history of our country, and it is for this reason that I have suggested, in my book *Down the Bay*, that Tōtaranui would be an appropriate name for this national park, perhaps mitigating in part the displacement that the title 'Abel Tasman' makes local iwi feel. [8]

The tī kōuka is one of the conspicuous trees on the coastal flat at Tōtaranui and several large examples are clustered near the mouth of Tōtaranui Stream. It is unclear whether these are actually native to the area or were planted by Māori or farming settlers. One was planted on a pā south of Awaroa Head. Like tōtara, tī kōuka are rare in the national park owing to the dominance of acidic soils derived from granite. However, in at least one place, a wetland bordering the northern side of Awaroa Inlet, a natural stand grows with kahikatea. This same type of vegetation once covered part of Tōtaranui.

Like tōtara, tī kōuka are extremely useful trees. Tī kōuka leaves are exceptionally strong and also durable in seawater, hence they were used in the construction of fishing nets, as well as kete to collect seafood. They were woven whole or as dressed fibre for ropes such as anchor ropes, and were used to construct tough footwear — pāraerae. Some cloaks were woven from the leaves, too, and individual fibres were used as nooses in waka kererū, the tōtara water troughs set in trees to attract and snare the birds.

Tī kōuka was also used as a vegetable, with the growing tips of the branches snapped off, trimmed and steamed. In fact, this use was copied by Captain James Cook and other early European sailors, hence the name 'cabbage' tree. In many places tī was actually cultivated, or at least encouraged to grow by clearing off competing vegetation. The trees were also harvested for the underground stems, which are rich in sugar, a use that ancestral Māori brought with them from Polynesia. A related species, tī ngahere, is common in moist open places, such as former slips, along the tracks in the park. Its leaves are even stronger than tī kōuka leaves. There is a very important sandbank in the park named Onetahuti, literally meaning 'burn tī on the beach' (although usually mispronounced 'One-ta-huti'), where local iwi believe dry tī ngahere leaves were used to make smoke signals.

Of great historical significance in the park are groves of karaka trees that occur near former pā sites (or have now spread into the regenerating bush), such as those at Taupō Point, Te Matau Separation Point and Tonga Island. Karaka is one of the 'tropical' trees of Aotearoa; its place of origin is Northland, but it has been transported to offshore islands and south as far as Banks Peninsula. It was planted for its edible seeds. The seed inside the beautiful orange fruit is poisonous and required prolonged cooking in a hāngī, followed by leaching in water for several days. But once prepared it is a nutritious source of carbohydrate, protein and fat, and would have been an important winter food.

At Tōtaranui a single old karaka grows alongside the main creek. While it looks like a remnant of a former Māori grove, it was actually planted by a member of the Pratt family, who purchased Tōtaranui in 1892. The tree was planted in a glasshouse built by William and Betsy Gibbs, the first farmers there, and although it survived it was trimmed to prevent it damaging the glass. As a result, the canopy today is low and misshapen. A key factor that supports this history is that the nuts of the tree are small (2 centimetres long), whereas Māori selected and planted varieties that had larger seeds (3–4 centimetres) and therefore produced more food.

Tōtaranui also has a historic avenue of plane

trees, planted by the Gibbs family in 1856. They form a marvellous entrance to the park's headquarters, and have become colonised by thousands of pirita, a species otherwise rare in the park. Introduced historic trees create an important feature of the park, including old pine trees planted as shelter by farming settlers. There is one at Goat Bay. If karaka trees planted by Māori in the national park have historic significance, then so too do those planted by early Pākehā.

M āori burnt the coastal bush in order to encourage the growth of rārahu, harvesting its aruhe, which contain starch (and cyanide). Aruhe was a staple food available more or less anywhere at any time, although it was favoured when harvested from key fertile sites and in early summer, when its starch content is high. A dense cover of fern delays the regeneration of bush, and even today there are both large expanses and small patches of rārahu on some of the sunny slopes of the park. Pākehā settlers repeatedly burnt off the rārahu to encourage grass, so it was eliminated over large areas.

When the settler farms were abandoned after less than a century, the grass was gradually replaced by kānuka, which now grows extensively through the park. In the natural environment kānuka would have been very much more restricted to small open sites in the beech forest created by windthrow, fungal infection or insect attack. Kānuka also colonises stable sandspits. It is a fast-growing species that establishes in very dense, pure stands, and this density makes the trees grow tall and straight. It is also a hardwood and can grow in hot, dry, infertile soil, so is ideal for any exposed slopes of granite.

In 1989, anthropologist Rod Wallace conducted a classic archaeological study to identify the trees used for wooden Māori artefacts held in selected museums. This found that kānuka was used for patu aruhe, hoe, weapons, spades and, especially, kō and other agricultural tools. All these items require reliably tough wood. The relatively short life of kānuka, plus its shallow root system, means that the trees often fell and died or were washed away by floods and wave surges. Then, kānuka firewood would have been locally plentiful. Collecting firewood was a daily necessity for Māori settlements and kānuka was also a major source of firewood for Pākehā settlers. Imagine all those wood-burning stoves but no pine or eucalypts. It is little wonder that today kānuka firewood is the most expensive on the market.

These days, the repeated burning or cutting of kānuka has ceased and kānuka forest has become a habitat. Its bark and small-leaved canopy host myriad insects and, therefore, birds, and periodically the canopy is clothed in a snowy dusting of flowers. Their nectar and pollen are major sources of food, especially for native bees. One of the really important features of kānuka is that it supports mycorrhizal fungi around its roots, which means it is a great habitat for orchids. The fungi help the trees, and orchids absorb nutrients from otherwise infertile soil. One of the latter is the hidden spider orchid, which lacks chlorophyll and flowers underground, so few people ever see it. Another is hūperei, the black orchid, 'perei' referring to the gem-like quality of the flowers and 'hū' being short for uwhi, the edible yam that was introduced by the Polynesian settlers. Hūperei has a branched system of fleshy underground stems that are filled with starch and were harvested as food. Unfortunately, wild pigs have discovered hūperei in the park and root up and destroy many of the plants, so there is a serious local conservation problem.

Finally, kānuka is also a nursery species. Pākehā logged rimu intensively throughout the park, but it is now widely regenerating under kānuka. In moist areas ponga form a dense understorey. When these grow up and form a trunk, rātā can colonise as epiphytes, and in this way rātā forest is gradually returning to the park. So, kānuka, once repeatedly burnt as 'scrub', has become a major ecological asset.

A similar story unfolds around mānuka. It was part of the wetland vegetation but also colonised infertile areas when bush was burnt, and it can form a natural low canopy tree in the stunted upland forest on acidic granite soil. It is a smaller tree than kānuka but also grows in extremely dense stands. Māori harvested mānuka for a similar range of items as kānuka, but mānuka was even more important for agricultural tools because the wood is exceptionally strong and durable in the ground. In addition, the young stems were used to line the bases of waka.

Mānuka was also well known by Māori for its health and healing properties. An infusion of the bark was used externally and internally as a sedative, and to treat scalds and burns. The ash from the bark was rubbed onto skin to treat skin diseases, while vapour from leaves boiled in water was used for colds. The inner bark was boiled and the liquid used as a mouthwash. Mānuka gained the English name tea-tree because of its value as an infusion for sailors and early Pākehā settlers. However, it was largely despised, like kānuka, because of its ability to invade farmland.

In Northland, mānuka is known as kahikātoa for its pink flowers (pink or red was the colour of kahika, or chiefs), and this is the origin of a wide range of beautiful horticultural forms. Mānuka 'tomato stakes' are a traditional garden support. The resin-rich wood is durable and aromatic, and popular as chips for smoking meat.

In modern times mānuka leaves are steamed and distilled for mānuka oil. Flavesone, leptospermone, isoleptospermone and grandiflorone are naturally occurring organic compounds, collectively known as triketones, which have very high antimicrobial properties. Mānuka varieties high in triketones have been propagated and planted in plantations for harvesting. Mānuka honey has also become sought after for its health value and, therefore, commands a high price. The usual antimicrobial agent in honey is peroxide, but a superior compound, methylglyoxal, is present in high concentrations in mānuka honey.

Mānuka is not the only tree in the park that is used in the honey industry. There are many hives seasonally using rātā, kāmahi, kānuka and honeydew (the product of scale insects on tawhai rauriki). The high numbers of introduced wasps are a threat to this industry, although, given that nectar is a food source of native birds such as korimako, and formerly of kākā, you have to wonder whether having hives in a national park is appropriate in any case.

Just a short way along our track we cross a stream, and here is an entirely different set of species that depend on high moisture and light. They include small trees like tutu, makomako, kōtukutuku, kanono, kawakawa and rangiora. All of these species once had their uses. Tutu was a source of sweet fruit juice, used to flavour the bland aruhe, although elaborate precautions were essential to remove the poisonous seeds. The species is a nitrogen fixer, can grow in sterile soils along slips, streams and, today, roadsides, and is a valuable conservation plant with a massive biomass turnover. Kōtukutuku was a source of edible berries called kōnini, much admired for jam by Pākehā settlers, while kanono has bright yellow inner bark that was used as a dye for colouring flax fibre.

Kawakawa, named after the Pacific kava and with the English name pepper tree, was and is a highly valued medicinal plant used to cure wounds, diseases, and digestive and kidney issues, as well as being revered for ceremonial roles. Its active ingredient is myristicin, which is a naturally occurring insecticide. The compound is also found in anise, dill, nutmeg and parsley, and occurs in the essential oils of these aromatic plants. For this same property, mānuka was used in kūmara beds to ward off damaging insects.

Today, plants like mānuka and kawakawa are grown commercially, and their essential oils distilled and used in a wide variety of applications, including as supplements and skin preparations. Totarol, the resinous agent in tōtara wood that makes it durable, is also used in a similar range of products. These uses reflect a growing interest in natural remedies from unique Aotearoa plants.

Tree ferns grow magnificently along the Abel Tasman Coast Track, especially the majestic mamaku, whose cooked pith was a source of food for Māori. These huge plants produce masses of debris that, in time, supports the regeneration of long-lived forest trees such as beech and podocarps. The fibrous stems of whekī were once valued as building materials by both Māori and farming settlers, while the ponga was used to mark trails in the dark. Aotearoa became internationally renowned for its spectacular tree ferns, in particular the ponga, which became the country's emblem. The silver lining under the leaves, unique in the world of tree ferns, helps to protect the plant from desiccation and enables the species to occupy a wide range of habitats, contributing greatly to the regeneration of forest after disturbance.

B eech is the dominant forest in the park, all five native species being present. It was an important food source for Māori in the form of kiore, the Polynesian rat, which came to the forest in huge numbers during mast (heavy) years of beech seeding. Today, kiore have been replaced by ship rats, a serious problem for native birds and snails. In traditional Māori culture, however, ingenious rat traps were set along trails and the cooked rats were stored in pātua made from the thick antimicrobial inner parchment of tōtara bark.

Beech forest was also a useful source of fungi, one of which is puku tawai. This is a large spongy bracket fungus that grows on decaying hard beech trees. Puku tawai often crash to the ground and lie in soggy masses, but if they are collected and dried they make a remarkable tinder. The tissue burns very slowly and can be put out only when smothered. Māori carried these smouldering tissues around to kindle fires, and the early Pākehā settlers learned about them and called them punk. An edible fungus found along the track in the park's beech forest is the native shiitake.

Beech bark was used for black dye and also for tanning, because doing so improves the lifespan of fibres and canes — including kareao — in the sea. Harvesting the bark for tanning was an important industry, giving the name to Bark Bay. Beech trees would have formed a major component of coastal driftwood coming down the larger rivers and spilling into the sea from coastal slips, providing a major source of firewood. Early Pākehā used beech a great deal, splitting the trunks for fencing and pit-sawing them into planks for housing and boatbuilding.

Today, beech is notorious for its mast seeding, which happens every few years and causes a spike in numbers of rodents, followed by stoats, and consequently a decline in birds. Hence, the beech cycle is an important element in the location, timing and intensity of anti-predator poisoning operations. Beech forest is the primary factor influencing soil and water conservation in the park, and the purity of the streams flowing from it results in high numbers and diversity of invertebrates, such as mayflies, and consequently excellent habitat for native fish and even whio. Like kānuka, beech forest is mycorrhizal, and the high diversity of orchids and fungi found here relates to this. It creates the most important ecosystem in the biodiversity and ecological functioning of the park.

The fires that destroyed most of the bush around Tōtaranui did not extend into the large gully systems leading into Goat Bay. Here, original bush survives today, including some of the largest rātā and rimu in the park, their trunks often smothered in a dense cover of climbing kiekie. Kiekie is a pandan, a group of tī kōuka-like plants that are widespread in the western Pacific tropics, where one genus, *Freycinetia*, is a climbing plant. Like nīkau and tree ferns, kiekie imparts a tropical luxuriance to the Aotearoa bush.

The long leaves of kiekie were and are

important weaving textiles, while the long roots descending the trunks of host trees made very strong binding fibres that were used to attach adze handles and the foot plate on kō. The flowers are borne in clusters surrounded by fleshy, sweet bracts, which were relished as a food, as were the ripe fleshy fruit. Pollination and seed dispersal are thought to have once been effected mainly by bats, but as these mammals have become rare it is thought that perhaps possums now carry out these tasks. Tree ferns are important hosts for the establishment of kiekie seedlings.

Large buttressed pukatea grow in the moist gullies and around wetlands. The bark of this beautiful tree was used medicinally. The inner bark contains an alkaloid called pukateine, which is a painkiller with properties similar to morphine and was used to treat nerve disorders, skin ailments and toothache. Pukatea was a name for Astrolabe Roadstead, a stretch of water used by French explorer Jules Dumont d'Urville in 1827, and probably refers to the fact that this sheltered coast was used to repair waka damaged in the rough seas of the West Coast and in Cook Strait. Pukatea wood was an important timber because it is light in weight but soft and not at all brittle. Often the carved tauihu was made from pukatea for this reason because it could take a physical pounding without being damaged. It is a beautiful timber with a soft brown colour and attractive grain well suited to joinery, and was sometimes known to early Pākehā bushmen as 'bastard rimu'.

Growing with pukatea in the park is nīkau. The tip of the palm was a valuable food source but harvesting it killed the tree and therefore it was used only in times of famine. More important were the huge fronds, which were used to line the roofs of dwellings. In addition, the red fruits attracted kererū, another food source. Nīkau is one of the most loved trees in Aotearoa, creating luxuriant gullies and groves 'marching' across hillsides.

Goat Bay stretches south from Skinner Point, a rocky headland named after Gerry Skinner, Member of Parliament for Motueka from 1938 to 1946, who strongly supported the creation of the national park. There is a lone tōtara there, a sentinel for the golden sand of Tōtaranui. Goat Bay itself, however, is bordered by a sand flat. When stable, this habitat carries a forest type uncommon in the park, notably many tītoki and tarata, which require fertile and well-drained soil, and along the coastal front there are large, spreading ngaio trees. The leaves of ngaio are punctuated with glands that produce an antiseptic oil, able to 'draw out' septic sores and toxic to sandflies and other insects — it was even once used as a sheep dip. Ngaio has become an important shelter-belt tree in rural coastal farmland.

Tītoki is one of the most beautiful and interesting trees in Aotearoa. It is a classic 'tropical' tree, one of 30 or so members of the *Alectryon* genus, ranging from South-East Asia and Australia to Aotearoa, with the majority in Queensland and New Guinea. The name *Alectryon* derives from the Greek word for rooster, referring to the glandular-looking, bright red flesh covering the seed. The contrast between the red flesh and the shining black seed is startling, and clearly attractive to birds such as tūī. The seed is poisonous but also contains a quality oil. This was expressed from the crushed seeds by Māori, mixed with aromatic leaves and used to anoint the skin. The timber is very strong and was favoured by Pākehā settlers for tool handles and wheels.

Akeake is a relative of tītoki and named for its extremely strong, heavy wood, which lasts 'forever and ever'. It was used to fashion tools and weapons such as taiaha. This remarkable tree is distributed virtually worldwide thanks to its papery seed capsules, which contain air and can float on seawater for months. The species evolved in Australia and has spread around the world over the last couple of million years — not enough time, seemingly, for it to evolve into separate species. Its resistance to salt spray has led to its valuable role in coastal shelter belts and the re-establishment of coastal forest.

Tarata is an extremely attractive tree, with large, shiny leaves and a white-barked trunk. Its common English name is lemonwood, which refers to the aromatic yellowish-green leaves. The species' traditional value lay in its scented oil, found in the leaves and bark. It is exuded when the bark is cut and soon solidifies into a seal. The gum was mixed with tītoki and other oils as a body lotion, and the gum was chewed as a mouth freshener — such as after a meal of dried shark. The essential oil contains nonane, which in plants is an uncommon inflammable hydrocarbon and a component in kerosene.

This richness of valuable species in the park's coastal forest illustrates a general point: it is very important to understand the habitat requirements of different species so that they can be managed in a sustainable way.

There are many other tree species along the Abel Tasman Coast Track, and herbs and shrubs are found there, too. When they first encountered them, people from Polynesia would have immediately applied what they already knew and recognised, and they would have embarked on a journey of experimentation. Either the plant looked like one from home, or they gave it a new name based on its key features. Every plant would have been tested for its use as food, fibre and medicine, applying well-known methods to improve and maintain the harvest and the quality. Pākehā acted in exactly the same way, except they had Māori to teach them directly. For example, what was tītoki became known by Pākehā as New Zealand ash because the leaves looked similar to those of the European ash and the wood had similar uses.

Both cultures had a severe impact on the local ecology. Māori exploited the fauna and caused many extinctions, and they burnt a significant part of the bush. Pākehā cleared bush for farming, logged trees for timber, and introduced many weedy plants and pest animals. Both cultures emerged from early exploitive phases with a conservation ethic, and both have recorded much of the knowledge as mātauranga Māori and as science. Today, it is important that we keep this knowledge alive, and value nature in the Abel Tasman National Park both for itself and as a treasure trove for the future.

N M ADAMS

A Line
Between
Two Trees/
Observations
from the
Critical Zone

———

Anne
Noble

[T]o attend to a tree's song . . . is to
touch a stethoscope to the skin of a
landscape, to hear what stirs below.[1]

The title of this tree project, A Line Between Two
Trees/*Observations from the Critical Zone*, riffs off a
multidisciplinary science initiative that focuses on the
Critical Zone, a scientific descriptor for that living, breathing, near-surface layer of the Earth, from the earth of
a forest to its canopy, that is experiencing catastrophic,
largely invisible loss of biodiversity. Part lover of science,
part interrogator of science, I began this project with a
question to provoke the kind of speculative musing that sits
within the domain of art: Do trees talk? And if there were
such a thing as a language of the forest, how might we
hear it? What quality of attention might we bring to bear?

In his long-form poem *The Book of Questions*,
the Chilean poet Pablo Neruda poses questions of trees
that invite us to consider the rich nature of human curi-
osity: 'What did the tree learn from the earth / to be
able to talk with the sky?'[2] Neruda's book comprises a
collection of many such questions in the form of unan-
swerable propositions that all refuse the possibility of
rational answers. Instead, they invite us to think with the
attentive intuition and imagination that animates our

capacity for wonder, and a feeling for enchantment that is in direct defiance of logic or common sense.

In my work with bees I have been fascinated by the notion that the insects have language and that a hive, in turn, is a conscious entity reliant on a complex unity of aural and sensory communication systems.[3] I have listened to the tuneful hum of a healthy hive and observed bees conversing through touch, vibration and their wonderful dances. I have also heard the sound of a queenless hive, and felt it as a groaning body of bees, together humming a series of discordant lower tones. I have no doubt that a colony of bees maintains its complex interior world within the hive and its relationship to the world beyond through a language that we cannot understand through the lens of a human logocentric perspective.

And so to my question 'Do trees talk?', which I am posing with quasi-scientific seriousness. If the rational answer is, 'Of course they don't — trees don't have language', then how might I imagine it differently? What can I do to find out? Should I sit by a tree and hope to hear it whispering? Or should I investigate, inventing a way to measure the tenor of the conversations between trees?

While talking trees appear in literature and

films, these are flights of our fancy — anthropomorphised creatures given mouths that speak the language of human beings. Instead, I am adopting the notion of a tree language that is not framed with reference to human speech, but is a language we cannot recognise, hear or understand. I like to imagine a forest or the bush operating like a colony of bees, with its own complex, interconnected consciousness, engaged in aeons-long conversations that are way beyond human comprehension. While this work is a nod to the revelations of Canadian scientist and ecologist Suzanne Simard about the latticed fungal network that connects trees and distributes nutrients,[4] *Observations from the Critical Zone* has its source in my reflections on the nature of language itself and how we might imagine it as a faculty of all interconnected living systems.

In his famous essay 'On Language as Such, and on the Language of Man', German philosopher Walter Benjamin says, 'There is no event or thing in either animate or inanimate nature that does not in some way partake in language'.[5] Regarding language as an inherent attribute of all living and non-living things suggests in turn that it is an interconnecting force linking organisms, matter and phenomena within complex environmental systems.

Perhaps then trees *can* be heard, and their

language might be visualised as a strange inscribing of processes occurring over time that resound within and for a specific material community — like a forest. This language would be, in Benjamin's words, 'immediate and infinite like every linguistic communication . . . And it is magical (for there is also a magic of matter)'.[6] Considered today, Benjamin's language of things invites us to reframe how we imagine and see the collective worlds of bees and the forest, and the marvellous conversations inherent in them.

So, I buried a length of photographic film in this precious zone of the Earth — the Critical Zone — with the idea of tracing tree signatures, musing about the possibility of capturing some form of tree language through the conceit of burying a medium receptive to the passage of time and chemical signatures in the ground between two trees.

A LINE BETWEEN TWO TREES, 2018/2020
Observations from the Critical Zone I
Exposure 312 hours
Kodak Portra 160 ASA
Bundanon Estate, Illaroo, Australia
Detail from an original image
7325 mm X 427 mm

A LINE BETWEEN TWO TREES, 2018/2020
Observations from the Critical Zone II
Exposure 2,488 hours
Kodak Portra 160 ASA
Bundanon Estate, Illaroo, Australia
Detail from an original image
7325 mm X 427 mm

Among Trees, Among Kin

Kennedy
Warne

As soon as I open the gate, I am among them. Kahikatea. Scores of telephone-pole trunks rising to neck-craning heights, swaying in pendulum arcs in the warm spring wind. Beside them, pūriri, so different in form, branching broadly, standing solid and immovable. I pace out one branch and find it extends 18 metres from the trunk. The base of the trunk is rippled with the weight of wood it supports — and not just wood. The forks of some branches hold massive clumps of kahakaha, or perching lilies, living islands of foliage known to early European settlers as widow-makers: if they fell, they could crush a bushman working below.

I run my hand across the pale corky pūriri bark and push a finger into tunnel entrances made by pūriri moths. These splendid creatures, whose wings are a match for the green glory of kākāpō plumage, spend up to six years as larvae inside the tree before they emerge to mate, lay eggs and die, all in a matter of days. The bullet-holed trunks of these immense trees testify to all the inhabitants that have sojourned within, life within life.

A boardwalk, built to protect regenerating saplings from human feet, leads me deeper into the green interior. The regrowth is profuse. Springing up from the soil are young kohekohe, coprosmas, karaka. I pick a kawakawa leaf and chew it, letting its sharp pepperiness flood my tongue. Fluted *Griselinia* trunks as thick as my forearm twine like vines around host trees as if holding them in a botanical embrace. Kōtare in the canopy call their piercing, insistent kingfisher notes, while tūī make dazzling aerobatic flights through the obstacle course of trunks. They seem to do it for the pleasure of aerial mastery. Within minutes of being in this place, my spirit revives. In the embrace of green places, I cross a portal between isolated human and interconnected being. I experience the relatedness of whakapapa.

Yet this place is not Te Urewera or a national park or some other sublime expanse; it is a mere 3-hectare scrap of remnant forest in Auckland's densely built-up North Shore, named Smiths Bush after the former landowner. To reach it, I have passed cricketers honing their batting skills at the Takapuna Cricket Club, and on one boundary of the reserve I peer through a thin fringe of foliage to see cars streaming along the Northern Motorway. To the dismay of those who lobbied to have this place preserved, the motorway split Smiths Bush in two when it was built in 1959.

Perhaps twentieth-century roading engineers had a professional disdain for forest fragments deemed to be obstructing a desired route. The same insistence on taking State Highway 1 through the middle of a forest reserve occurred near Hunterville, disturbing a gem of ancient podocarp forest bequeathed to the country by Scottish settler and politician Robert Cunningham Bruce — the same man for whom the road to Ruapehu's Whakapapa skifield is named. In this case, the state highway was eventually rerouted. Today, if you walk into Bruce Park you follow what was once the route of the main highway and is now a path being reclaimed by grasses and woodland.

The greenness of Bruce Park makes me dizzy. Moss-coated kareao vines twine in wild green tangles. The trunks of rimu, mataī and tōtara are thickly velveted with moss and fern. Kererū flash green and white as they fly heavily overhead. Pīpīwharauroa sing their springtime song but hide themselves well, so I cannot see the banded chests of these shining cuckoos. I am alone here, and the place seems under a spell. I want to walk without breaking a single twig, to preserve the enchantment.

I come to a memorial erected in 1924, recording that the park was 'given to the people of New Zealand and to the residents of this district in particular, in order that they may have ever before them a beautiful specimen of New Zealand forest life'. It was a far-sighted and necessary gift. All around this 30-hectare enclave are pine plantations, farmland, scattered oaks, poplars and macrocarpas. The park is an island of indigeneity in a sea of introduced species.

I have a fondness for these fragments and remnants, and seek them out wherever I go. They are the 'last, loneliest, loveliest' leftovers (to quote English author Rudyard Kipling) in a country that was once almost all forest.[1] And such a diversity of forest. It occurred to me while walking in the mountains of Otago recently that if the New Zealand landmass had formed sideways, east to west, rather than north to south, the forests would have been profoundly similar. It is because the country spans so many degrees of latitude, and thus such a climate range, that there can be dry subtropical kauri forests in the far north and wet temperate podocarp forests in the deep south.

Mountain ranges and mountain rain shadow create other variations. I hold in my fingers a fallen tawhairaunui leaf. I have picked it up from a trail in Mount Aspiring National Park. In such alpine land-scapes beech leaves form confetti carpets across the forest floor. The red beech themselves are almost Japanese in form: dark trunks, spindly branches and a froth of small, shining leaves. The trunks are sometimes black with fungus nourished by honeydew, a nectar produced by scale insects feeding on beech sap. Many times I have touched the tip of my tongue to the tiny globes of honey-dew that stand out on threads from a beech trunk, to sip their sweetness.

W here podocarp forests are massive and moist, beech forests are delicate and airy — but no less ancient. Beech leaves have been found in Antarctica, when what is now the frozen continent was part of Gondwana, long before it moved to its polar address. To walk among beeches is to be transported to an ancient time, even as it is a transportation of aesthetic delight.

It is also a walk in sacred space. In Māori cosmology, all trees possess sacredness as the offspring of the spirit-being Tāne. The forest world is te wao nui a Tāne, the great forest of Tāne, and te wao tapu a Tāne, the sacred forest of Tāne. According to some traditions, Tāne presented the forest as a gift to his parents — Ranginui, the sky father, and Papatūānuku, the Earth mother — to ease the pain of their separation. Kahikatea — those sky-stabbing trunks I watched swaying in Smiths Bush — carry Earth's words of love to the uplifted sky.

Humans are part of this story, too. Like forests, we are the offspring of Tāne, but we are the younger sibling. Tāne made trees before making humans — a story not dissimilar to the first chapter of Genesis, or to the mapping of the human genome, which shows we share a third of our genes with oak trees. Trees are our kin. We share a sacred bond.

It is not difficult to sense the sacredness of trees when you stand in the presence of a rākau rangatira, a chiefly tree. I did so in Pureora Forest Park, west of Lake Taupō, when I visited a tōtara of such antiquity and esteem that it has been given a name: the Pouakani tōtara. Although the tree is signposted from the main road, the track to it is little used. It weaves agreeably through podocarp forest: broadleaved hardwoods and tree ferns, a forest type as primeval as beech.

I paused beside a young tōtara, perhaps 50 years old, that had a broad vertical gash in its bark stretching from ground level to above head height. This tree had been selected for future use as a waka. Such trees respond to having their bark cut by laying down extra wood on that side, creating a density of timber suitable for the bottom of a canoe hull. The process takes decades. Whoever sliced this tree was thinking generations ahead.

I walked on. About the time I thought I must be close, I glanced to the right of the track and thought for a moment I was seeing an outcrop of rock. Then my knees went weak and something buckled within me. It was Pouakani, and its presence was overwhelming. I walked around a fence that had been erected to protect the tree and looked up into its plant-drenched canopy. This rangatira tree began its life 1800 years ago, around the time that the Taupō volcano erupted and flattened the central North

Island. It precedes all human life in Aotearoa.

How Pouakani escaped the axe and chainsaw I do not know. Logging in Pureora Forest started in the 1940s, and by the late 1970s the New Zealand Forest Service was determined to clear-fell the remaining merchantable timber: mataī, rimu, tōtara, kahikatea — the pick of this country's lowland podocarp trees. This was despite the fact that in 1977 a 341,160-signature petition had been presented to Parliament demanding an end to native forest logging.

A group of conservationists from the newly formed Native Forest Action Council decided that something more emphatic than a petition was required. They drove to Pureora, obtained a camping permit and set up camp — but not on the ground. Using a few ropes, but relying mostly on vines, Tarzan style, they climbed into the canopy of six tōtara trees, built platforms and staged a treetop sit-in, declaring that Pureora must be spared.

It was a watershed moment, and it led not only to the protection of Pureora, but also to the preservation of another outstanding podocarp remnant, at Whirinaki, on the edge of Te Urewera, and eventually, in 2002, to the end of logging in state-owned native forests throughout the country. Had protesters not taken to the Pureora treetops in 1978, Pouakani and its forest community would surely be gone.

How should a person respond to a rākau rangatira? In Japan, a straw sash may be tied around a tree of this significance, a mark of respect for a vessel of spirit. In Thailand, trees have been ordained, their trunks wrapped in orange robes to signify sanctification. In Māori culture, certain trees were known as tipua, beings possessing supernatural powers. Passing such a tree, a traveller might pick a twig or fern and place it at the base, murmuring a karakia.

According to the Roman historian Aelian, the Persian king Xerxes the Great once saw a majestic plane tree during the course of a military march. He halted his army and ordered that they set up camp around the tree so that he could admire it, and paid homage to it by adorning its branches with necklaces and bracelets. Aelian records that when Xerxes marched on, he left a caretaker for the tree, 'as if it were a woman he loved'.[2]

American poet Mary Oliver had a touchingly direct way of relating to a beloved tree. Each year, on the first day of spring, she visited an ancient oak she had named Noah, to hug and kiss it, and to listen to its leaves tremble in reply. This she did for 30 years. In one of her poems Oliver wrote that being among trees inspired her to 'walk slowly, and bow often'.[3]

The illustrious ornithologist Charles Fleming once remarked that New Zealand's podocarp remnants are this nation's Gothic cathedrals. Being in the presence of a tree like the Pouakani tōtara is an inescapably numinous experience. It invites, at the very least, a disposition of deference, of bowing — or, following a more contemporary practice, taking a knee.

I learned something of Māori etiquette towards trees from a Tūhoe woman who lived in Ruatāhuna, in the heart of Te Urewera. Nanny Pohutu was well versed in rongoā, traditional Māori healing practices, and one winter morning she took me with her on a plant-collecting trip. I met her at a patch of forest near the local school. She had been in the forest from soon after daybreak. Dusk and dawn are when the mauri, the life force, of the forest is strongest, she told me.

We paused at the edge of the ngahere for a few moments, and Pohutu said a karakia, asking the forest to welcome us. Everything in the Tūhoe world starts with prayer. It is a matter of respect, an acknowledgement of human limitation, an orientation to the unseen world. We waited a little longer. 'The ngahere is like a marae, or someone's home,' Pohutu said. 'You wait to be invited in.'

We followed a sun-dappled path, and Pohutu pointed to this plant and that, stroking the leaves as she passed: kōkōmuka, a species of hebe, for curing rashes; manono, a coprosma, for cuts and sores; a type of fern for toothache; tāwiniwini to ease a cough; makomako for tired eyes; and kāramuramu, another coprosma, for detox. How did she know which of a tree's leaves to pick, I asked. 'Ask the tree,' she replied. 'Let the mauri within you connect to the mauri of the plant.'

Pohutu was collecting the leaves of patete, and there were many of these slender green shrubs lining the path. 'Perhaps one is moving more than another in the wind, and it catches your attention. Perhaps a bird is flying around one. You might go past and then turn back, because one stands out to you. There are many ways a connection is made.'

Pohutu sang a waiata as she picked. Her soft, lilting notes blended with the sounds of a stream that babbled beside the path. She told me she had grown up with some of this medicinal knowledge, absorbed through watching her mother, but paid scant attention back then. She finds it remarkable and humbling that so much has come back to her.

Now Pohutu herself is gone, her knowledge passed on to others, her demonstration of connection to the life of trees a vivid and memorable witness.

Human connection and kinship to trees is a principle common to Indigenous cultures. But evidence of connection *between* trees has been one of the great scientific discoveries of recent times — a mind-shifting revelation that trees are social creatures, linked intimately via the 'wood-wide web', a network of roots and fungal threads that enable trees to communicate with and nurture one another.

One of the pioneers of this work is Suzanne Simard, a professor of forest ecology at the University of British Columbia. She found that even unrelated species such as Douglas fir, an evergreen pine, and birches, deciduous nut-bearers, can share carbon via a fungal web that links their root systems.

Networking has been found in Aotearoa forests, too. In 2019, two Auckland researchers showed that kauri trees were sharing water and nutrients through an interconnected root system. The trees were 'holding hands' beneath the ground, they said.

This finding would not surprise Indigenous people. 'There's no such thing as a tree on its own,' Ngātiwai kaumātua Hori Parata told me when I asked him about the disease that has been killing kauri trees in recent years. Trees are part of a whānau, he said. 'When the whānau is intact, everything is healthy and strong. When it starts to break up, that's when you get things like this dieback disease.'

The wood-wide web has seized the imagination of many writers. Authors such as Peter Wohlleben (*The Hidden Life of Trees*), Peter Tompkins and Christopher Bird (*The Secret Life of Plants*), and Robin Wall Kimmerer (*Braiding Sweetgrass*) have all explored the evidence and implications of trees' socialising.[4]

Kimmerer, a Native American botanist who expertly and elegantly combines scientific thinking with Indigenous knowledge, writes about pecan, a mast-fruiting species that bears its nuts in synchrony, with fruiting episodes sometimes years apart and barren years in between. How and why these trees fruit when they do remain a mystery. 'If one tree fruits, they all fruit — there are no soloists,' she writes. 'Not one tree in a grove, but the whole grove; not one grove in the forest, but every grove; all across the county and all across the state. The trees act not as individuals, but somehow as a collective.'[5]

How are they doing this? Kimmerer points to the subterranean mycorrhizae — the fungal filaments that inhabit and interconnect tree roots. In redistributing carbohydrates from tree to tree, these networks act as a kind of Robin Hood, she writes. 'They take from the rich and give to the poor so that all the trees arrive at the same carbon surplus at the same time. They weave a web of reciprocity, of giving and taking. In this way, the trees all act as one because the fungi have connected them. Through unity, survival. All flourishing is mutual.'[6]

The same tantalising idea of reciprocity and interconnection as the central reality of forests, if not the entire natural world, is the theme of Richard Powers' masterly book *The Overstory*.[7] Powers compares the underground networking of trees by fungal intermediaries as a kind of massive public health system — social welfare for forests. 'Every sense we have that it's every individual for itself out there, a competition among ruthless individual forces, is completely laid waste to by this vision of massive interconnection,' he said in a 2019 conversation.[8]

Powers said that the book arose from a road to Damascus conversion he had — from being tree blind to tree obsessed. 'I couldn't tell an ash from an elm,' he confessed.[9] His aha moment came when he met his equivalent of the Pouakani tōtara: a redwood that was '30 feet across and 300 feet tall and almost as old as Jesus. I'm not proud of the fact that it took something this size to effect this conversion. But once the conversion was under way, everything changed. I went back out east to the forests I thought I knew, and the most banal thing was now a mystery.'

I would not say my own fondness for forests has been the result of a conversion as dramatic as the one Powers describes, rather a steady and joyous awakening. Twenty years ago, in my first story for *National Geographic* magazine, I wrote about walking through a Fiordland beech forest in the rain. With the cowl of my rain jacket pulled down over my head, I compared myself to a monk on his way to vespers. In my mind, I am still that monk, still on his way to a place of connection.

Kimmerer points out that the words 'humus' and 'human' arise from the same root. This is not surprising: the Genesis creation story describes the first human as being shaped from soil. Why should it not follow that humans might have an intrinsic sense of belonging to the forest world, with its deep, decomposing layers of organic matter, the origin and destination of so much terrestrial life?

Among trees I am among kin. It is both a mystery and a comfort. I revel in the mystery and dwell in the comfort, recalling poet Rainer Maria Rilke's amazed epiphany: 'because everything here apparently needs us, this fleeting world, which in some strange way keeps calling to us'.[10]

The
Golden
Bearing

———

Meredith
Robertshawe

T rees have been intimately connected with human lives since ancient times. Traditional myths throughout the world tell of trees in celestial and terrestrial realms that hold up the cosmos, create portals to unknown realms, and link the heavens, the Earth and the underworld, binding them together. Sacred trees have inspired pilgrimages, and enabled gods, heroes and ancestors to traverse times and worlds, and to seek enlightenment and immortality. Symbolising birth, death and rebirth, hallowed groves and trees have provided gateways for human connections with magical and natural worlds, and bestowed protection, knowledge, fertility — and light.

In the various Māori creation narratives of Aotearoa New Zealand, the universe was in a state of Te Pō, perpetual darkness, when Tāne-mahuta, god of the forests and one of many children of Ranginui, the sky father, and Papatūānuku, the Earth mother, wrenched his parents apart from their dark embrace. This action, or misdeed, allowed Te Ao Mārama, the world of light, to emerge through the darkness and bathe the newly separated Earth and sky — together with Tāne and all beings — in golden light.

An age and a world away, Virgil's ancient Latin epic poem the *Aeneid* describes the quest of Trojan hero Aeneas to visit his father in the underworld. The immortal prophetess Sibyl of Cumae informs Aeneas that he must first find 'the golden bough' growing concealed in a nearby forest to ensure his safe journey to this other world. Discovering the golden branch glittering among the leaves of a shadowy tree, the hero breaks it off and, favourably, another grows back instantaneously.[1]

Speaking about the genesis of his twenty-first-century work *The Golden Bearing* — a life-size sculpture of an archetypal tree-form covered in golden glitter — Aotearoa artist Reuben Paterson says, 'Trees, as ancient things, come with their own ancient histories, stories and genealogies.'[2] As the 2013 New Zealand artist-in-residence at New Plymouth's Govett-Brewster Art Gallery, Paterson (Ngāti Rangitihi, Ngāi Tūhoe, Scottish) found inspiration for, and connection to, *The Golden Bearing* through his Māori whakapapa, cultural heritage and creation narratives, and his father's work as a landscape gardener, alongside tree mythologies, art history, and the natural environment and designed gardens of Taranaki.

Paterson is of Ngāti Rangitihi, from the *Te Arawa* waka of Te Moana-a-Toi Bay of Plenty, and calls on his ancestry when talking about his art practice: 'Our kaumātua today still speak of our people as

"Te Heketanga-a-rangi" — those who descend from the heavens, in proud remembrance of the origin of our tūpuna, Ohomairangi, the son of Pūhaorangi, a powerful celestial deity and the protector of Rangiātea, as the source of learning and knowledge handed down by gods and spirit-ancestors from the 12 heavens.'

Māori astronomical knowledge encompasses the creation of the universe, celestial bodies, gods and light, and Paterson's whakapapa provides a direct connection from these ancestral beginnings to his use of glitter in his art practice. He talks of pure, glittering light as he recounts the first hara or misdeed of Tāne-mahuta, the placement of celestial beings and the coming of Te Ao Mārama, bringing mātauranga, whakapapa and tikanga, the bases of human existence.

After Tāne-mahuta separated the primeval parents Ranginui and Papatūānuku, the world was still in darkness. Tāne asked his siblings Tangotango and Wainui — the parents of the sun, the moon and the stars of the Milky Way — to give their children to him. Placing Te Whānau Mārama, this family of light, in three baskets, Tāne took them to the sky in a sacred waka, where they were tipped out, adorning the sky with light and illuminating the Earth.

Paterson envisages the brilliance of Te Whānau Mārama mapping celestial pathways for

Māori navigators across glistening ocean waters to the sparkling sands of New Zealand's coastlines. He sees this continuing through to his own work, which is inspired by this natural and spiritual world, and acknowledges the values of light and whakapapa through cultural readings of places and objects. 'I spent much of my childhood on the black sands of Piha, with that volcanic heat you feel right through your feet as you run to the sea; the curling purple rivulets streak the sand like vivid auroras. I lived in a landscape painting surrounded by its own naturally glittering surface.'

When Paterson arrived in Taranaki for his artist's residency, he was struck by the beauty and greenness of the region's public gardens and curated parks. Seeing these gardens as a potential site for his work, he recalled what he refers to as his 'seedling years', when as a young boy he was growing up along the road from his late father's very first landscaped gardens — planted with saplings that have now grown into luxuriant trees. Occasionally returning to this neighbourhood, either in person or through childhood memories, Paterson considers the growth of the trees to be a record of his father's hands digging the soil, planting the seeds and creating a designed landscape. He regards it as a 'real spectacle

Reuben Paterson, *The Golden Bearing*, 2014,
Pukekura Park, Ngāmotu New Plymouth.
*Photograph Bryan James, courtesy
of Govett-Brewster Art Gallery.*

of time . . . moving on in a natural cycle, for each yearly ring of growth I, too, grow into each year of his passing'.

Through his work, Paterson invites us to look twice at the landscapes we occupy. *The Golden Bearing* was initially exhibited not in an ancient forest or a mythical realm, but rather in Pukekura Park and and Pukeiti, two of Taranaki's most celebrated gardens.[3] In these places, he says, the sculpture was 'a Tāne, linked back to the first light'. It acted as a focal point that invited people to reflect on their relationships with nature, allowing visitors to navigate through, and bear the weight of, the 'undulating rhythms, visual truths, energy and history' that landscapes hold.

Although these two landscaped gardens retain areas of indigenous bush, they are curated in part within European and English cultural traditions, in which gardens were composed akin to classical paintings, framed as subjects for viewing. This idealised landscape painting aesthetic — inspired by, but not true to, nature — was popularised by the depictions of ancient historical scenes and landscapes by seventeenth-century French baroque painters Nicolas Poussin and Claude Lorrain. These influenced the next generations of artists and gave rise to the use of the Claude glass. When 'scene-hunting' — the Victorian pastime of finding views to visit and paint — artists would literally turn their backs on their selected landscapes and, by looking into their dark-tinted Claude pocket-glasses, like glancing in a rear-view mirror, paint the flattened view reflected in the slightly convex glass.

Abstracted from the surroundings and with simplified colour and tonal ranges, this way of looking led artists and viewers to see land and country as picturesque scenes-to-be-painted. English Romantic 'painter of light' J. M. W. Turner was among those influenced by these artists and their style. In 1834 he painted *The Golden Bough*, depicting a golden-light-filled and dream-like landscape scene from Virgil's *Aeneid*, in which the Sibyl of Cumae holds the mythical 'bough of radiant gold' aloft in the light before entering the underworld with Aeneas.

Paterson's siting of *The Golden Bearing* — a 4.5-metre-high golden sculpture of a tree supplanted incongruously in an idyllic gardenscape — had the effect of three-dimensionally exploding this symmetrical composition, creating, he says, 'a painting, but in the real'. While acknowledging the principles of the golden ratio and concepts of ideal beauty and form espoused by the ancient Greek philosopher Plato and the Italian Renaissance architect Leon Alberti, *The Golden Bearing* simultaneously replicates and deconstructs the aesthetics of landscape painting. In this constructed environment, the contained lawn set the foreground, mature trees were thrown into full background relief, and diverging waterways and paths provided lines of perspective. The scene was, to borrow Italian writer Umberto Eco's words, 'at once absolutely realistic and absolutely fantastic'.[4]

At both Pukekura Park and Pukeiti, *The Golden Bearing* looked like a tree and performed like a tree, moving in the wind. When it was first installed, some visitors thought it was a real tree that had been painted gold and covered in glitter. There was some outrage. 'How could you do this to a *tree*?' People formed connections with the tree, and wanted to protect it and step into action for trees. However, the component materials of *The Golden Bearing*, essential as they are to the physical survival of the work, are employed to withstand, rather than be part of, nature.

Constructed in situ on a concrete base, the work comprises a metal skeleton supporting a reinforced polystyrene and fibreglass trunk form, with more than 500 plastic, wire and fabric leaf sprays fixed to the tree's extended branches, creating a canopy 4 metres in diameter. The sculpture is adorned with gold spray paint and thousands upon thousands of tiny hexagonal and square microplastic shards, 50 kilograms of glitter held in place with two-pot primer, glue and automotive clear coat.

It's the golden glitter that enables visitors to

experience *The Golden Bearing* not as the result of toxic products and processes of industrial manufacture, but rather as something unearthly and beautiful. This abstract, impenetrable, glittery surface effect at once creates and obfuscates the idea of a 'tree', making the very air around *The Golden Bearing* shimmer.

Paterson has worked with glitter since his time studying at the University of Auckland's Elam School of Fine Arts and considers light to be an activator of his work. The daily cycles of celestial bodies — the sun, moon and stars — transform *The Golden Bearing* through a complex interplay of light, and the tree constantly reflects its environment. Monochromatic gold elevates the sculpture from the representational to the fantastical, and Paterson harks back to the use of alchemy, turning base materials into gold, imbuing them with qualities of magic, power, wealth and the Divine, and aligning them with the sun as the radiant centre of our universe.

Paterson sees gold as also representing the transitory colour of autumn, remarking on the 'performance' of two tall redwoods with golden leaves that take centre stage each year in his own garden, 'the seasonal drama unfolding day by day'. Just as Paterson is attracted to the autumnal performance of his golden redwoods, so visitors were drawn to the theatrical nature of *The Golden Bearing*. Presented during the summer seasons, at the time of New Plymouth's annual Festival of Lights and World of Music, Arts and Dance (WOMAD) events in Pukekura Park, the sculpture attracted fairy tea parties, picnics, storytelling and dress-ups.

The work's theatricality was heightened further when *The Golden Bearing* was later exhibited within the Govett-Brewster Art Gallery.[5] The sculpture glitters brightly in the sunlight — more brightly than any real tree — but it also fades, and faded glitter does not stand up to harsh gallery lights. Not only that, but without a summer breeze to breathe gently through its leaves, the sculpture becomes static. In the gallery, this most artificially constructed of environments, the tree seemed — strangely — more unnatural than in the gardens and had to be made hyper-real. A shiny new layer of glitter and a hidden fan softly rustling the leaves ensured that, to borrow from Eco once more, this 'faked nature corresponds . . . to our daydream demands'.[6]

People asked gallery staff, 'Is it *real?*' Children decided, 'It *is* real!' Well, no and yes — *The Golden Bearing* is arbo-real. It's a real artwork made by a real artist, and it stirs real emotions and responses. Paterson enjoyed the fact that *The Golden Bearing* was community centred and could be encountered by passers-by. He looked forward to seeing the role the tree would take on in the lives of Taranaki people: 'I loved watching how children interacted with the tree, discovering it. As adults, we often forget the magic that makes living so interesting — we forget we can still look at things for the first time.'

Acknowledging *The Golden Bearing* as an illusion intensified visitors' attraction to it. They blogged about it and Instagrammed selfies with it — standing with their backs to the tree, holding their smartphones aloft to re-create the perfect view, showing their personal connections with the tree and redeploying these images in digital space through the Claude glass of our times. The accompanying social media campaign elicited countless photographs and responses. Some Instagram posts got right under the surface of the sculpture — 'real gold naa just fake'. Other blogs and posts imagined something otherworldly.

Like Aeneas, people souvenired parts of *The Golden Bearing*, plucking leaves from its golden boughs and exposing the brown and green plastic fabric underneath. Echoing the mythical restoration of the *Aeneid*'s own golden bough, new golden leaves grew every day on the sculpture as Govett-Brewster Art Gallery staff repaired the work, colluding with Paterson to present — in spite of its fallibility — a perfect, gilded 'tree' and offering Eco's 'absolute unreality . . . as real presence'.[7]

The social media posts showed that *The Golden Bearing*'s visitors did, in fact, see this golden sculptural tree and other, real, trees anew. They contemplated the fact that trees are sacred in Māori culture and something precious to be protected. They decided they preferred natural trees — 'wouldn't swap this for a real tree!' — and yet they appreciated the stimulus of the sculpture in leaving them 'wondering about how we connect and relate to nature', while walking among trees and 'turning off the thoughts and soaking up the quiet and green'.

As real trees — the natural lungs of the Earth — provide oxygen for us to breathe so *The Golden Bearing* lets our imaginations breathe. Paterson is an observer of the symbiotic relationship between humans and trees: 'It is people that give *The Golden Bearing* life, that somehow let it come to life, becoming a real tree,' he says. His gilded sculpture merges Māori and European ancient narratives with contemporary ideas, dislocating our perceptions of nature and of art. *The Golden Bearing* asks us to look twice, to reflect and to consider — what is the essential treeness of a tree?

The Peculiar
Trees of
Aotearoa

Glyn
Church

Perhaps you need to be a foreigner to appreciate how peculiar the trees of Aotearoa New Zealand really are. If you grow up with our native forests, they're simply familiar, and anything you see on a regular basis becomes normal. But our trees are far from normal, and by world standards they are bizarre. Having spent more than 40 years in this country, I'm still flabbergasted by how strange and different the native flora is compared with that of other temperate climates. This country is one of three 'botanical arks' — floating islands with hundreds of unique plants. Along with Madagascar and New Caledonia, we have some of the highest rates of endemism, or plants that are unique to this land. Out of a total 2300 different species, roughly 80 per cent of our plants are endemic.

Forty years ago, I was, in part, inspired to come to Aotearoa by Joseph Banks' sojourn here. Banks was probably the most famous man in England in the 1770s and regarded then as the intrepid explorer of the South Pacific, with Captain James Cook as his chauffeur. Banks and his companion Daniel Solander were the first Europeans to botanise in this magical land.

Like Banks, I came to Aotearoa specifically for the plants and trees. I intended to stay just two or three years and then return to the United Kingdom as a southern hemisphere plant expert. Banks had studied botany at the Chelsea Physic Garden on the edge of the Thames in London, and I also worked and studied there, in what was then a 300-year-old garden, for three years in the mid-1970s. At that time Banks was virtually unknown in the United Kingdom — unlike in Aotearoa, where he's still famous. Worldwide, Cook is far more famous now than he was back in 1771, but in his day, he was merely the pilot of a boat.

At just 1.6 hectares, the Chelsea Physic Garden is a small garden, yet it contains approximately 5000 different plants, and for its size has probably the greatest botanical diversity of any garden in the United Kingdom. It was during my time at the garden that I made another significant connection with Aotearoa plants. I had recently graduated from a three-year study course at Pershore College of Horticulture in Worcestershire in the heart of England.

Two of my fellow students went off to work at the Botanischer Garten München-Nymphenburg (Munich Botanical Garden) in Bavaria, Germany, and it seemed an ideal opportunity to visit. It was the early days of cheap flights, and my wife Gail and I would fly from London to Munich for weekends.

During one of our many visits to the Munich Botanical Garden we came across a greenhouse displaying stupendous tree ferns. I vaguely knew that tree ferns had lived on Earth during the dinosaur era and that they had become part of the coal and oil measures beneath the ground, but I had no idea that they still existed — and yet here they were in all their primeval glory right in front of my eyes. I'm told I was like a kid in a toy shop, running around in an excited fashion and leaping from label to label to discover where these plants originated. On the plane back to London I began planning our move to Aotearoa to see these fabulous trees in the flesh.

It wasn't an easy journey. The authorities didn't need anyone with Gail's training and experience in primary school teaching, and they certainly didn't want horticulturists. And to make matters worse, a visa wouldn't be granted without a job, yet no one would give you a job unless you had a visa. But for some reason the authorities *did* grant us an interview at New Zealand House on St David's Day, which was a promising sign as we are both Welsh and David is the patron saint of Wales.

For reasons we've never fathomed, we were given visas — and not just for the two years we had asked for, but for permanent residency. We were on our way to the land of tree ferns and podocarps. I had studied podocarps in my botany classes at Pershore, and at the time I had wondered why I bothered as I was sure I'd never see such trees in my lifetime. More on them later.

We arrived in Aotearoa on Anzac Day 1976, and within a short time I was working for the New Plymouth Parks Department. I came here knowing hardly a single native plant, such was their scarcity in the United Kingdom at the time, and I spent those first three years gobbling up as much information as I could about the native flora. Everyone at the parks, from the director down to the youngest apprentice, was keen on plants and I found myself in the perfect learning situation. I spent my weekends tramping, trying to come to grips with this strange land and its even stranger plants.

For instance, I knew South Africa had proteas, and I knew about the closely related Australian proteas — called *Banksia*, after Joseph Banks — but it was a shock to find that the world's tallest proteas grow here. Rewarewa forms a perfect pillar, as if clipped by shears, and in springtime it produces velvet-textured, rusty-red flowers. These long, pointy flowers, braced like coat hangers, finally burst open and the petals coil like a spring. Rewarewa is so peculiar that it instantly became my favourite tree — after the tree ferns and podocarps, of course.

O ver the years I've done my best to share the wonders of New Zealand's plants with others. I did this first while working for the parks department, by offering seeds of native plants to botanic gardens around the world. Chelsea Physic Garden and Hortus Botanicus Leiden (Leiden Botanic Garden) in the Netherlands were the first two gardens in the world to exchange seeds, starting way back in 1683.[1] Gradually, more and more botanic gardens joined the scheme, so that today Chelsea Physic Garden exchanges seeds with more than 350 gardens and universities around the world. These exchanges between botanic gardens are free of charge and the scheme is free of politics.

Because New Zealand's native plants are so amazing, I enquired if any gardens here were in the scheme and sending seeds overseas. What I discovered horrified me. It appeared that our major public gardens were using the scheme to acquire seeds but were not sending anything in return. I decided to change that single-handedly, and in 1977 began collecting seeds in order to put out my own list. After a short time, I had a list of more than 100 different seed types gathered from a mixture of natives and exotics.

For 17 years I put out a seed list, known as an *index seminum*, and sent it to a variety of botanic gardens around the world. Each year I would choose more gardens to add, and through this I found that Communist countries were keener on receiving seeds than Western countries, with the exception of France and the United States.

Chinese gardens were especially eager to receive my seeds, and I was swapping seeds and letters with Kunming Botanical Garden long before New Plymouth and Kunming became sister cities. I received many kind letters promoting long-lasting friendship, especially from Beijing in China, Leipzig in the former East Germany and Tallinn in the former USSR. (Following their independence from Russia in 1991, the Estonians were so poor and yet so proud of their new status that they unglued printed envelopes with their old USSR address and reversed the paper so they could print their new Estonian address on the outside.)

Because I didn't have a garden of my own, I rarely asked other gardens for seeds, even though each sent me its list every year. My only motive was to share the very rare commodity of southern hemisphere seeds around the world. When Gail and I bought a 4-hectare property in 1986, however, I began asking for seeds and imported some new and unique plants, an activity that was perfectly legal back then.

As we transformed our new Taranaki property from grazing to trees, we created our own arboretum. It was timely in many ways, as I was invited that same year to join the International Dendrology Society, an organisation dedicated to enhancing our knowledge of trees. In order to become a member you either have to own an arboretum that is open to the public or be responsible for looking after a significant tree collection, as well as being nominated by two current members. As our tree collection has matured, we have hosted dozens of overseas botanists and dendrologists, most of whom have been very impressed by our rare specimens, many of which were imported as seed from botanic gardens back in the 1980s and early 1990s. They include such rarities as *Acer serrulatum*, *Ailanthus vilmoriniana*, *Alangium platanifolium*, *Alnus ferdinandi-coburgii* and *Aphananthe aspera* — and that's just some of the 'A's.

Plant experts from China, South Korea, Germany, France, Austria, the Netherlands, the United States and the United Kingdom have all made special pilgrimages to view our collection. I've also travelled extensively with the International Dendrology Society to look at forests and trees in other countries, and each time I'm reminded of just how unusual New Zealand's forests are. They are evergreen when, really, they should be deciduous given our latitude, and much of our bush looks like tangled tropical jungle rather than uncluttered temperate forest.

It's easy to take our forests for granted, with the view that that's just the way they were made, but for overseas plants-people they defy belief. If we compare our forests with those of the eastern seaboard of the United States, for example, it's clear that they are the complete antithesis. Our forests are all evergreen and largely have a jungle-like appearance, whereas the forests of the Appalachian mountain chain, extending from Canada to Georgia, have an open, airy feel. The northern Canadian end is very cold, but in Georgia the climate is steamy and much hotter than anywhere in Aotearoa. And yet the forests of Georgia are deciduous and have the same tree species as those in New York state or Canada.

Botanists from Georgia can scarcely believe that our forests are evergreen and yet our climate is cool temperate. This mystery deepens when we take altitude into consideration; even up in the mountains of the Southern Alps, the Tararua Range and Mount Ruapehu, our forests are still evergreen despite the much cooler conditions. Georgian botanists say their forest is deciduous because they get a few frosts every winter, but the winter temperatures at North Egmont on Mount Taranaki or at the ski field at Mount Ruapehu are much colder than they experience.

Much of our forest is made up of conifers and hardwoods, the latter being broadleaved trees such as beech and kāmahi. The podocarps, like the tree ferns, are relics of the dinosaur era and are peculiar because they are conifers that lack cones — which is, of course, a contradiction in terms. *Podocarpus* means 'foot fruit', and the trees produce 'fruits' rather than cones. These are reminiscent of plums and berries, and are created by the stems becoming fleshy and, in some cases, enveloping the seeds. Podocarps are found in Pacific and Indian Ocean countries, and Aotearoa hosts one of the main centres of the family, including rimu, kahikatea, miro, mataī and tōtara.

The understorey of our mixed hardwood-podocarp forests is full of creepers, climbers, ferns, mosses and epiphytes, creating a mesh of vines and greenery that is difficult, if not impossible, to penetrate. By comparison, when you step into a European or Appalachian forest it's possible to walk in virtually any direction. Nearly all of our conifers, with the exception of the remarkable kauri (and a few cedars), are podocarps. Kauri forests are also virtually impenetrable and include other tree species, often with a tropical feel such as taraire and toru. Our kauri is enormous compared with the kauri of New Caledonia and Vanuatu, which are home to numerous small species.

Beech trees dominate many of our forests and, in comparison, these seem uncomplicated, with a more open understorey. However, even they do not conform to international norms. Chile and Argentina are at a similar latitude to Aotearoa, and like us they have many species of beech tree. The difference is that while their beech trees are nearly all deciduous, ours are all evergreen — even those that grow near the tops of our high mountains.

What's even more remarkable is that many of the evergreen trees that grow in our coastal forests are remnants of a tropical era. Coastal regions of Aotearoa have a Pacific or equatorial flora, as many of our native trees have close relatives that grow much closer to the equator. Our representatives of these genera are the outliers, the ones thriving in the coolest climate.

These 'tropical-looking' trees are predominantly found in our coastal regions where the temperatures are more equitable. They include karaka, pūriri, pukatea, tītoki and tawa. Another characteristic linking them to the tropics is their long, pointy leaves, with a pronounced 'drip tip' — a sharp tip that allows the leaf to shed water quickly, so that it doesn't stay wet and rot in a constantly wet climate.

Some of our trees — notably pukatea — display the tropical trait of buttress roots, which help hold the trees steady in unstable swampy land. Pukatea, kahikatea and maire tawake growing in wet soils have breathing roots, called pneumatophores, and northern rātā and pōhutukawa use their aerial roots to grow in air without soil. For instance, the largest pōhutukawa forest in the world is on Rangitoto Island, where, it is believed, the original trees grew 'in air' while attached to pure lava rock. Other tropical features of some of our native trees are cauliflory, where flowers appear on the trunks and branches (as in kohekohe), and ramiflory, where flowers grow out of woody stems (as in kōtukutuku and māhoe).

T his brings me to the topic of violets, giant daisies and fuchsias. The botanical name for māhoe is *Melicytus ramiflorus*, the species name referring to the characteristic of ramiflory. An even more remarkable feature of this tree is that it's probably the largest violet in the world. Violets and pansies are usually pretty little plants, smaller than ankle height, so how did some become trees?

We're also famous overseas for our tree daisies *Senecio*, *Brachyglottis* and *Olearia*. Daisies are generally thought of as lawn weeds or ankle-height flowers, or sometimes growing up to 2 metres tall, as in Michaelmas daisies. In Aotearoa, however, we have a host of native tree daisies, most of which are woody shrubs or, some would argue, trees. The *Fuchsia* genus of plants includes around 100 species, most of which are at knee to waist height. Aotearoa is represented by four species, one of which, kōtukutuku, is the biggest in the world, growing to 15 metres. This magnificent species has a hefty trunk and attractive orange-brown peeling bark.

The eccentricity of our native flora is further illustrated by plants that pass through a series of juvenile phases. This characteristic can be recognised by the emergence of different coloured leaves at different stages of life, peculiar leaf shapes corresponding to various stages of life or, to add to the confusion, divaricating or tangled phases when the tree is young. Many of our young native trees start life with brown or red leaves that gradually turn green as the tree gains height and maturity. Examples of these include kauri, rimu, mataī and various members of the lancewood or *Pseudopanax* genus.

Several theories for this last habit abound; one suggestion is that it made the trees less visible to grazing moa because the colour of the living tree blends with the dead leaves on the ground. A more likely scenario is that the red pigments protect the young plants from intense sunlight, which could damage tender tissues. In New Caledonia, the new growth of many shrubs is an intense red that protects the tender leaves from strong sunlight in the absence of overhead forests.

The ability of our native trees to produce different leaf shapes for each stage of their lives is yet another conundrum. Very few countries have plants with such a pronounced display of juvenility in their young trees, and early European botanists visiting Aotearoa often gave two or three different names to one species observed at different stages of growth because they figured they were different plants. Two classic cases for potential confusion are horoeka or lancewood, and toothed or fierce lancewood, both of which can assume five or more distinct leaf changes during their lifetime. A juvenile leaf of horoeka is typically 50–60 centimetres long and around 1 centimetre wide, but at maturity it more usually measures 15 centimetres by 3 centimetres. Furthermore, if the tree is cut, the new leaves will revert to juvenile growth. A stump of horoeka can remember the type and size of leaf it has had at each stage of its life.

There are many trees in Aotearoa with distinct juvenile phases, including matai, with lacy brown leaves that are completely different from the rich green adult foliage; pōhutukawa, with thin, delicate, shiny leaves that are unlike the thick, tough adult leaves; and pōkākā, which starts life with tiny battleship-grey leaves that expand to finger-size dark green leaves at maturity.

Several theories have been proposed as to why

many of our shrubs and a few trees — including kōwhai, mataī and pōkākā — have a juvenile divaricating trait, which is rare elsewhere in the world. One theory suggests that it's to protect the inner part of the plant against cold, another argues that divarication might confuse browsing animals such as moa, and yet another proposes that divarication might protect trees against dry conditions or wind. As Aotearoa is one of the windiest countries on the globe, that may be a clue.

What happened to my quest for those tree ferns that so attracted me to up sticks and move to Aotearoa? Well, I still love and admire them, and I can't imagine living in a place where they don't grow. However, while they seem very common in some regions, they are far from ubiquitous, and vast tracts of this country have none at all. Under the worldwide Convention on International Trade in Endangered Species of Wild Fauna and Flora (CITES) agreement, most tree ferns are regarded as endangered. As such, it's illegal everywhere to sell a tree fern from one country to another unless it has been grown from spores in a nursery. Our native *Cyathea* tree ferns are thus protected.

A local exception to the CITES regulations applies when removing a pine plantation, in which case a special permit can be arranged to uplift *Dicksonia* tree ferns before the pines are logged, and the removal of *Dicksonia* for fibre is permitted in some regions. While tree ferns are generally regarded in Aotearoa as being so prevalent that they can be used to create 'ponga' walls, I wince every time I see such wanton waste. Tree ferns are precious and likely to become scarce as time goes on. We should preserve them whenever possible.

Other remarkable trees in this South Pacific haven include nīkau and tī kōuka, both of which are not strictly trees but rather giant monocotyledons — a group of plants that we usually associate with grasses and lilies. Our native nīkau is a thing of beauty and adds a very 'tropical' flavour to our coastal regions. It grows naturally as far south as Banks Peninsula and Wharekauri Chatham Islands, and has the distinction among palms of growing at the highest latitude.

Our iconic tī kōuka is also called 'cabbage' tree because it is edible, albeit very bitter. As temperatures warm in the northern hemisphere, this strange plant has graduated from being a rare specimen found only on the south coast of the United Kingdom, where it is known as the Torquay (or Torbay) palm, to be grown universally across England and Wales as an exotic addition to local gardens.

How, with all these peculiar features, might the trees of Aotearoa help us solve some of the environmental challenges facing us now and in the future? Which of our trees should we be planting to meet the demands of climate change? Three or four examples spring immediately to mind. Species such as pōhutukawa, with its remarkable tolerance to saltwater inundation and salt winds, and mangroves, which trap and bind silt and sediment and reduce the impact of waves, might provide our first lines of defence against rising sea levels.

Kāpuka has proved to be highly resistant to wind at sea level (hence its scientific name *Griselinia littoralis*), and for this reason is grown in the Channel Islands of Jersey and Guernsey to protect crops from the roaring Atlantic gales. It is an obvious contender for native forest regeneration. Mānuka has undergone a total makeover in the time I've lived in Aotearoa, going from an 'invasive' scourge, which the government and farmers tried to eliminate, to flavour of the decade. Apart from the bonus of therapeutic mānuka honey, the plant itself is an excellent nurse plant for re-establishing native forests. It achieves this by protecting delicate native tree seedlings from the elements and through its ability to grow on impoverished soils, making it perfect as the initial stage of revegetation.

A request from three researchers in California to meet up with me a few years back when they visited

Aotearoa, and the discussion we subsequently shared, caused me to rethink my understanding of climate change and our arboreal world. Having made a study of ancient redwoods by climbing the trees and measuring every aspect of them on an annual basis for more than 20 years, Steve Sillett, Marie Antoine and Robert Van Pelt discovered that each new year in a tree's life is the best year ever for creating heartwood.[2] This has been a revelation.

At ground level we might watch young trees grow at a rapid pace and imagine that they are super-charged carbon sequesterers, but in reality it is our oldest trees that create most new heartwood growth each year and, therefore, are best at carbon sequestration. Old trees, it seems, are doing the greatest job of all, and while planting many new trees is obviously a good move to capture carbon, old trees may be much more important in this respect. Sillett, Antoine and Van Pelt are changing the way we view old-growth trees, and other research-ers worldwide are finding similar results with other tree species.[3] Together, these results present a compelling argument for putting much more effort into saving the native forests we already have.

Another exposure to international thinking has influenced my thoughts on forest regeneration. While we talk about re-creating native forests on poor hill country here, we could learn much from South Korea. On a visit there in 1998, I spoke with the Minister of Forests and several Korean botanists, who explained that during the occupation of their country by Japan between 1910 and 1945, the Korean forests were decimated and the timber was taken to Japan.

Once the country was released from the yoke of war in 1953, thousands of South Koreans toiled for free for their nation at weekends to build railways and wharves and plant trees on the barren hills. As you travel through South Korea today, it becomes apparent that many of the forest zones comprise trees of the same height and girth, all planted by those willing volunteers.

Korean culture esteems resilience and regeneration, and this is seen in both the replanting of forests in South Korea and in their choice of national flower, *Hibiscus syriacus*, which is virtually indestructible. Despite chopping, crushing or burning, the bush is able to revive, and symbolises South Korea's own regeneration.

When I asked if the hand-planted forests were suitable for logging, the minister replied that they were not for milling but would remain there for ever. He explained that South Korea buys all the timber it needs from the Philippines and Aotearoa. He saw his role as custodian of the nation's forests. It is a refreshing point of view.

Tree Sense
of Place

Jacky Bowring

N.M.ADAMS

W hat are those intangible, emotional bonds that tie us to place? Even in the face of disruption, change and threats, our connections to place can form aspects of identity and belonging. Sense of place is increasingly being recognised as a significant component of people's health and well-being.[1]

The significance of place is seen most intensely when people are removed from their familiar landscapes, or when those landscapes are threatened or changed. Changes to the landscape can induce feelings of power-lessness and loss. Some changes are the direct result of human actions, such as mining and urban sprawl. Other transformations result less directly from human actions, such as the impacts of climate change through increasing risk of wildfires, floods and extreme weather. And nature itself is also a change-maker, with natural disasters such as earthquakes dramatically transforming places.

The dislocation of people from place is sometimes termed 'root shock', described by American social psychiatrist Mindy Fullilove as 'the traumatic stress reaction to the destruction of all or part of one's emotional ecosystem. It has important parallels to the physiological shock experienced by a person who, as a result of injury, suddenly loses massive amounts of fluids.'[2] The idea of root shock makes a direct connection between people and trees, the distress of people echoing the shock experienced by the roots of plants when they are transplanted and when the changed conditions can be severely damaging. The resonances between humans and trees are fundamental to a 'tree sense of place'.

T rees are a vital component of sense of place, for they 'gather places around themselves'.[3] Their roles as part of place range from acting as structural elements — behaving like walls and thresholds — through to being human-like characters and storytellers. The longevity of trees embeds them deeply into feelings about place; they become, as English writer Fraser Harrison puts it, 'landmarks, milestones and other points of reference by which each person can take his or her own bearings in time and place'.[4]

The value of trees is made most vivid when they are threatened or removed, when all of the roles they play suddenly become even more apparent. Greg Moore, chair of the National Trusts of Australia Register of Significant Trees Victoria committee, describes how trees are core to our relationships to place, and that 'people tend to understand that they're important, even if they haven't articulated or rationalised their attitude'. He adds, 'Trees are markers and totems in our lives and in our landscapes — most of us have some sort of connection with them.'[5]

There are many examples of the way in which the significance of trees to sense of place is made tangible. In Christchurch's post-earthquake residential red zone, for example, the trees structure and enfold the sense of place, even after the buildings have gone. Other types of natural disaster, such as fires and tornadoes, can mean the complete loss of trees, and their absence speaks volumes about their significance for orientation and for grounding in place. Felling — or even the threat of felling — also intensifies our affection for them.

The role trees play in sense of place is, in part, related to the way they fulfil specific functions: to structure space, as something to climb, to shelter under and, importantly, to provide food. Christchurch's 600-hectare residential red zone is a clear illustration of this. The red zone is the result of the demolition of more than 7000 homes following the series of earthquakes that began in 2010, with the most damaging on 22 February 2011. The soil in the red zone is prone to liquefaction, and so large areas of the city's eastern suburbs were ruled no longer suitable for housing. The removal of the homes here created an enormous green park, and intense debates over how the land should be used followed.

The houses were cleared, but the trees remained.

They have since been recognised as having important roles; many of them are survivors from domestic gardens, planted for their productive value. An interactive 'smart map' plots the location of a wide range of trees, including fruit and nut trees, and foraging has become a common practice in the city.[6] In recognition of the value of these red zone survivors as food producers, olive, apple, walnut, quince, pear, apricot, almond and peach are among the publicly accessible trees identified on the smart map.

The functional contribution trees make to place is also spatial, as landmarks and shelter. With the removal of built elements following earthquake damage in Christchurch, trees took on the roles of buildings. Artist Holly Best has observed that 'Trees are shelter and refuge; they reveal our desire for ownership, privacy, territory.'[7] In Christchurch's city centre, too, the trees endured the quakes, while around them 80 per cent of the buildings collapsed or were later demolished as a consequence of the damage they had sustained.

Comparisons were made with cities in war-torn areas — even London after the Blitz — but the defining difference in Christchurch was the trees.[8] Unlike many natural disasters, earthquakes do not significantly affect trees. In 2011, the year of the most damaging earthquakes, a heavy snowfall in July, rather than the earthquakes themselves, caused the loss of trees in the Christchurch Botanic Gardens.[9]

With the buildings in the city centre gone, strange new vistas opened up. City residents found themselves disoriented by the loss of familiar buildings as landmarks and began searching for street signs in order to navigate; trees began to emerge as significant new markers. Stoically holding their places within the urban landscape, the trees were familiar waypoints.

The remarkable Poplar Crescent was now much more visible, and it seemed to have grown in stature. In Latimer Square, which had been the base for search and rescue teams from around the world during the disaster response, the trees provided a kind of open-air, built form, defining the edges of the spaces and creating large outdoor 'rooms'. The trees along the Ōtakaro Avon River provided a green spine that could be seen from afar, providing a pathway and point of reference. Tī kōuka were traditionally used by Ngāi Tahu to navigate through the landscape, and in the post-quake city trees emerged again as an enduring wayfinding system.

Two powerful metaphors contribute to the role of trees in sense of place — trees as buildings, and trees as humans. The metaphor of trees as buildings draws upon their spatial qualities; as architectural proxies trees have a long legacy, extending back to the myth of French architectural philosopher Marc-Antoine Laugier's Primitive Hut as the origins of building. In 1755, Laugier, who was a Jesuit priest, described how 'Some fallen branches in the forest are the right material for his purpose; he chooses four of the strongest, raises them upright and arranges them in a square; across their top he lays four other branches; on these he hoists from two sides yet another row of branches which, inclining towards each other, meet at their highest point. He then covers this kind of roof with leaves so closely packed that neither sun nor rain can penetrate. Thus, man is housed.'[10]

In Christchurch's red zone the individual sections, or lots, remain marked by boundary plantings, and the grain and scale of humans living in place is very legible. Like Laugier's concept of a building made from trees, here are whole suburbs constructed only with vegetation: paired trees, lines of trees, tree corridors and tree archways, all representing the ways in which trees are metaphorically similar to buildings, constructing a sense of place.

Trees as humans — as *us* — is one of the most ancient analogies. Trees and humans share temporal and figurative qualities. The life cycles of humans and trees echo each another in a temporal sense, and we sometimes use trees as a datum for our own lives, moving slowly through to maturity. Figuratively, trees and humans both have trunks and limbs, and the resonances are sometimes uncanny, but even beyond these metaphorical dimensions are the comparisons of the characters and temperaments of trees with those of humans.

Japanese master carpenter Tsunekazu Nishioka praised a 2000-year-old cypress tree for its response to place. The tree had responded to its dry conditions by driving its roots down through the rock to get water, and 'It was precisely because of these harsh conditions that it had lived for 2000 years. It is the same with human beings. If they are indulged . . . given anything they want, they do not turn out well. Trees are like human beings.'[11]

British sculptor Andy Goldsworthy echoes this shared connection between humans and trees, recounting: 'One of the most powerful images I have of New York was staying in a hotel on Broadway. My room was high up in the building, I think on the 17th floor. I looked out of the window of my room and I saw a tree that had seeded itself, growing out of the side of the building opposite. It was for me a potent image of nature's ability to grow, even in the most difficult circumstances.'[12]

The metaphorical image of the resilience of Jews during the Holocaust was the inspiration for

Goldsworthy's memorial at the Museum of Jewish Heritage in New York, where he planted trees within hollowed-out boulders.

Trees are sometimes conceptualised as observers or witnesses of the landscape, emphasising their sentience. As journalist Will Harvie has written of his wanderings through Christchurch's abandoned and vacant eastern suburbs, the trees have witnessed the area's evolution: 'The trees were the great gift of the red zone. Off Fleete Street, in Dallington, a proud pair of willows dominated the landscape. Nearby, a delight-fully pruned conifer was witness to decades of care by somebody. Way north in Brooklands, gum trees were so large they must have been planted long ago by farmers. In Kaiapoi, an elm shimmered lime green in a pleasing sunset and slight breeze. Everywhere were clumps of cabbage trees.'[13]

Trees are containers of meaning; they hold the narratives of place. Across the red zone, the patterns of trees tell small, intimate stories of the streets and neighbourhoods: avenues, twinned trees marking gateways, the trees of churchyards where the church has gone. Some trees have what look like the graves of pets at their bases.

A blog on the suburb of Dallington, one of the areas most affected by red-zoning, recalls the cedar tree that stands on the grounds of what was St Paul's church and school. It now stands alone, and the writer, anthropologist Annette Wilkes, comments how the tree is: 'A place to stop, sit and contemplate. This iconic tree is important to the St Paul's community and to Dallington as a whole. To me it represents a tangible link to our past, to the many thousands of people in our community who spent parts of their lives on this site, learning, celebrating, spending time together. It remains to anchor our memories when the bricks and mortar of buildings cannot.'[14]

The interwoven narratives of people, place and trees are emphasised by the trees at St Paul's, which include a legacy of the 1980s, when the school's pupils 'were amongst those who celebrated Arbor Day with tree plantings. Many trees surviving on this site were planted and tended by young children who are still watching them grow.'[15]

Wilkes, a former resident of Dallington, is one of the so-called Quake Outcasts who battled authorities and insurers over the red-zoning of her property, which she felt was unjustified. Her family had deep roots in Dallington, going back to 1884, and the place is part of her identity. In particular, she had close bonds with two memorial trees: one marked the place where her dog was buried and the other, a tōtara, was a memorial to her father.[16] Wilkes was an active voice in trying to retain the trees as an important part of the red zone's identity.[17] She noted that her perspective as a social anthropologist meant weighing the trees in a very different way to the red zone arborists, who were evaluating them according to their stature and proportions as they made their decisions about which trees would be retained.

The potency of trees as storytellers is also recognised by artist and writer Simon Palenski, who described a visit to his father's red-zoned property in the east of Christchurch: 'Standing out in the open, where our yard used to be, he gave me a rundown of each tree and when he had planted it: paperbark maple, tarata lemonwood, houhere lacebark (his pride and joy) . . . When my parents left this house, they took from their garden whatever they could dig up and replant at the new one. His trees had to stay.'[18]

For her Master's research into how the experience of the earthquakes impacted residents' sense of 'home', Kelli Campbell carried out 29 interviews. Participant 24 in Campbell's research describes the trees in the red zone as 'markers', and how it was the trees that maintained a sense of connection, a tethering to place, once the house was gone: 'My dad lived five houses down and . . . he has a big karaka or something and so his house has gone now . . . all that is left is that [tree] and that is the reference point now and there is no other geographical reference point now. There is just that big tree, and me and my sister were even talking about it the other day that we are going to have to realise that the two cabbage trees are the fence line [to his home] on that side and the beech tree if it still stays there is . . . where the garage was. And you know everything else will be gone.'[19]

And Campbell's Participant 18 recalls: 'Hopefully when they get rid of all the other houses [in the red zone] they will just smooth the land out around the trees that were there and make it a big open reserve and let people wander through and go "Oh, my apple blossom is still there". . . I had an apple blossom and it was beautiful . . . [W]e got it when it was small and it cost us a lot of money and I am hoping when they knock the house down that they don't kill the apple blossom and the other trees . . . [W]e had in around our place some beautiful rhododendrons and camellias and things and they are getting really big and I hope that they would leave them there. Just remove the fences and then let the people of the town enjoy some of those expensive trees that people had in their yards. It would be lovely to wander through.'[20]

Attachment to place is fundamentally about an emotional bond, and bound up with this is the role trees have in the sense of place. The emotional bonds and affection for trees are vividly displayed when people are wrenched from their homes and from their tree sense of place, or when the trees are threatened. The loss of trees in North Carolina during Hurricane Hugo in 1989, for example, was devastating. A local journalist wrote, 'Next day, driving around town was like going to a friend's funeral', and others echoed that it was like the 'death of someone close'.[21]

In Christchurch's residential red zone, plaques have been placed on trees by those who used to live there. These markers are moving, emotional illustrations of how important it is for people to feel rooted in place and how they try to hold on to those roots even when they are displaced.

Our high emotional investment in trees often comes to the fore when they are endangered. Following the signing of a new road-maintenance contract in 2012, up to 36,000 mature street trees in Sheffield, United Kingdom, were threatened with felling. There were protests by locals, and some were arrested. The episode was examined by Matthew Adams, a psychologist who writes on environmental and social issues, who suggested that the fate of the trees illustrated a blunt financial approach by the authorities: 'Chopping down trees and replanting saplings elsewhere is apparently a lot cheaper than looking after mature trees and working around them.'[22]

The threat to the Sheffield trees revealed the long-held and deep attachments locals had to them. Their powerlessness to stop the felling was intensified by emotions of loss and grief, and they compiled 'tree stories' to express their feelings. One campaigner, Joanna Dobson, described street trees as 'quotidian landmarks', but those same familiar and ordinary quotidian qualities can mean that trees can easily be taken for granted. 'To rip them out, as our council is doing, is to destroy not only the tree, but also something profoundly important to the identity of our city and to those of us who call Sheffield home,' Dobson wrote.[23]

The loss of trees in a landscape can erase the sense of place. When a hurricane in 2011 wiped out the urban forest in Joplin, Missouri, it was the end of 'the everyday landscape that refreshed residents on a daily basis and contributed to a local sense of place'.[24] When a tornado hit Goderich, Ontario, in the same year, many of the mature trees that helped make Goderich 'Canada's prettiest town' were lost. Locals spoke of how devastating that loss was, how 'it does something to your soul, to be in this place when it's so smashed. And part of the thing that is missing is that — it was the trees.'[25]

Groups soon mobilised to replant them — to 'self-advocate, initiate, and accomplish as much as we can to heal our landscape'.[26] Trees were seen as a vital part of a 'reorientation' strategy that responded to the disorientation within the town and the damaged sense of place and well-being their absence caused. As testament to the significance of the therapeutic aspect of trees and of place attachment, geographers Amber Silver and Jason Grek-Martin have explained how, in Goderich, 'the monument erected to commemorate the one-year anniversary of the disaster was cast in the form of a broken tree showing signs of regrowth'.[27]

Greening is a recognised therapeutic aspect of rebuilding after disasters. Both the physical experience of working with nature and the mental benefits of doing so underlie the ways in which trees can help us get over things. The research of anthroecologist Keith Tidball illustrates how the act of planting brings positive outcomes and can re-establish place. Following Hurricane Katrina in 2005, the trees of New Orleans came to hold strong symbolic significance 'contributing to identity and sense of place'.[28]

The trees of Christchurch's residential red

zone have offered their own quiet therapy, steadfastly maintaining the pattern of the cultural landscape now that there are hundreds of hectares of vacant land, their familiar forms offering some solace. In other red-zoned areas of the city, trees have played more active roles in the well-being of residents, in some cases protecting their houses from rockfalls during the earthquakes or providing shelter in the aftermath.

T he intertwined nature of trees and place ties an emotional knot that binds us to landscapes. Sense of place is the rooting of meaning within a location, and with that our sense of familiarity, identity and belonging. Trees are a vital marker within this. As geographers Paul Cloke and Owain Jones have explained, 'Trees can construct places and vice versa. Many of the attributes of trees form common currencies in our understandings and appreciation of place; their size, rich materiality, their interconnectivity, their life cycles and seasonal cycles all offer qualities which are readily and vividly drawn into . . . concepts of place'.[29] This construction of place includes the functional, metaphorical, narrative, emotional and therapeutic roles that trees play.

While decisions are made about what to do with the land, the fate of the red zone trees in Christchurch remains in the balance. Planning for the zone's future involves a range of scenarios. Holly Best has described how 'Trees where they naturally belong are growing strong but are vulnerable to the city's Plan — waiting to be bestowed with the plastic tape allowing them to stay'.[30]

In Sheffield, 5500 trees had been felled by 2019, when the programme was halted in the face of the protests. Writing about the event later, Matthew Adams observed that 'Collective resistance and articulation of loss and melancholy also appear to have revived local-ised, creative expressions of love of place and reminded residents of the value they place on trees and our more-than-human companions.'[31]

Tree sense of place embraces how trees contrib-ute to making places — how they become indispensable parts of our familiar environments and the objects of our affections. Trees offer practical, functional contri-butions to the landscape, they feed us and they shelter us. Storytellers, they embed the narratives of place as markers and totems. Trees resonate with our sense of place, our feelings of home and even with our selves.

Part Two

—

Greening
the
Anthropocene

N.M.A.

Burying
the Axe
and the
Fire-stick

———

Susette
Goldsmith

Opening the Dominion Conference on Bush Preservation and Amenity Tree Planting at Parliament on 2 April 1937, Governor-General Viscount Galway told the 100-plus delegates that one of its main objectives was to develop a '"tree sense" in the people from one end of the country to the other'.[1] Furthermore, he said, planting roadsides, 'beautifying' the landscape and restoring 'as far as possible the one-time forest beauty of the country by preserving the existing native trees and by replanting others when the opportunity offers' were national concerns. Galway urged property owners to plant their roadside verges and to observe Arbor Day, and added that belts of 'centenary trees', with nameplates recording the names of the benefactors, would become 'fine monuments' and 'objects of beauty'.

The conference was partly prompted by Galway himself, who had appealed to the government as part of his own campaign to plant more trees throughout the country. The previous year, the *Evening Post* had reported that Galway would present a cup to the local authority that planted 'the best and most artistic mile of highway'.[2] This call to council action did not go unheeded. The Wairewa County Council, on Banks Peninsula, for example, budgeted £200 for roadside exotic tree planting or bush preservation to mark the upcoming centennial of the signing of the Treaty of Waitangi in 1840, and

Hāwera's mayor, J. E. Campbell, planned a 5-kilometre roadside avenue of trees between Hāwera and Normanby.

Minister of Internal Affairs Bill Parry had called the conference in an attempt to establish a national policy that would protect the country's indigenous forest. He chose his audience and his words carefully. Members of interest groups and government departments were there, along with promoters of scenery preservation and beautification, sawmillers, forest company officials, and Minister of Lands and Commissioner of State Forests Frank Langstone. Although they followed different ambitions, Parry pointed out, they were all working towards the same generally commendable end.

Despite all their efforts, however, they were losing the battle. Parry's interpretation of the 'tree sense' Aotearoa New Zealand needed was far more matter-of-fact than Galway's. What would be achieved by planting trees along roadsides, in parks and in odd corners if people continued to be careless of their heritage?, he asked the delegates. Every summer, whole hillsides were devastated by fires caused by carelessly thrown matches or cigarette butts, and land continued to be cleared of native bush for agricultural and pastoral purposes while other land already cleared, and of much greater value for farming, was left to grow weeds and put to no practical purpose, he added.

Parry outlined two proposals. First, the public needed to be educated to put an end to the reproach that 'every New Zealander was born with an axe and a fire-stick in his hands'.[3] And second, the various concerned organisations should be coordinated under a national body and a uniform policy implemented by local branches and committees. The country, Parry informed the delegates, was suffering from serious floods; stripped hillsides, previously covered in native bush, had slipped away in a night, ruining thousands of acres of rich land and causing river erosion at terrible cost to the country. In addition, he said, New Zealanders needed to get over the 'vexed question of natives versus exotics'.

This was a hot topic. Robust discussions on whether native or exotic trees were best suited for plant-ing in parks and reserves and for growing in state forests had crowded the letters to the editor columns of the nation's newspapers throughout the 1930s. Angry corre-spondents fed up with the devastation of indigenous forest that had gone on for many years advocated a mora-torium on the felling of native trees; others supported the replacement of native trees with fast-growing exotics.

The debate had been stimulated by a 1928 visit to the country by Arthur William Hill, director of the Royal Botanic Gardens, Kew, who addressed the government about what he saw as an indiscriminate planting of exotic trees in natural reserves. Reserves, he claimed, 'should be sacred to the native flora'.[4] The *New Zealand Herald* report on the address added that Hill's warning came at a time when community opinion in favour of planting only native trees was already strong and steadily growing.

Parry's solution to the 'vexed question', he told the conference, fell between the two extremes and included some further cutting of native timber alongside the preservation in perpetuity of selected forest remnants and planting of exotics on readily available land. Langstone and representatives of the Public Works Department, the Main Highways Board, the Post and Telegraph Department and the Railways Department put their own cases for what they considered to be appropriate preservation of trees.

Suitably briefed, the delegates went on to discuss the future of areas of existing bush, amenity planting in general, the purpose of Arbor Day, and the powers of local authorities and public bodies to plant and care for trees. The government should set up an interim committee to draft a constitution, they concluded, and a national body should coordinate the work of interested groups and educate the public. In the meantime, there was the prospect of war to be faced and the centennial to prepare for.

Historian Jock Phillips points out that rather than being a celebration of the signing of the Treaty of Waitangi, the 1940 centennial unofficially and predominantly became a 'tribute to the noble pioneers' who had conquered the land and paved the way for material progress.[5] Centennial processions — of which there were many throughout the country, often in conjunction with agricultural and pastoral association shows — routinely led with bullock drays, axes and pit saws, and concluded with parades of cars and tractors.

Viewed through the lens of native bush preservation, as advocated by Parry in 1937, these parades of tree-felling implements take on a more potent meaning: they are the villains of the piece. Their part in denuding the land, causing erosion, slips and floods, had contributed to the disastrous economic consequences Parry described to the conference delegates. The material progress and technological advancements in communications, railways, farming and industry that were extolled by the centennial film *One Hundred Crowded Years* were largely achieved at the expense of New Zealand's trees.[6] The 37,000-plus lights that illuminated the Centennial Exhibition buildings at Rongotai in Wellington might have shone a beacon on the nation's achievement of domestic comfort courtesy of electricity, but at the same time, and from another perspective, they represented the ongoing struggle the Royal New Zealand Institute of Horticulture was having with electric power boards and their 'unaesthetic onslaughts' on trees throughout Aotearoa as they installed posts and lines.[7]

In a letter to botanist Harry Allan, Parry's undersecretary, Joe Heenan, wrote electric power boards were the 'champion vandals of the whole country', who deliberately took power lines over trees in order to find an excuse to chop them down.[8] In 1934 similar objections to the rampant march of development had been made to the government by the institute regarding the wholesale destruction of native bush to make way for roads, railways and highways. The progress celebrated at the 1940 centennial had been fatal to the good health of New Zealand's indigenous forests.

Further reading about the centennial celebrations reveals some other ironies. The one-shilling centennial stamp, designed by Leonard Mitchell and titled 'A Giant Kauri', features Tāne Mahuta rising majestically out of Waipoua Forest, while the four-penny stamp, designed by James Berry and titled 'The Progress of Transport', shows ships, a train and an aeroplane, and bullock carts passing along a rough track carved through the forest.

No emblem more fitting to symbolise New Zealand's history could be found than the giant kauri, the *New Zealand Centennial News* informed its readers in 1939: 'It has witnessed the development of New Zealand, a modern Dominion of the British Empire, from a land unknown; it has seen bush and swamp turned to rich pasture; it has looked unmoving on the trials and triumphs of a young country elevating itself to nationhood.'[9]

If the tree had been able to speak, Tāne Mahuta might have had something different to say. At the time, public and political pressure to preserve the Waipoua kauri, halt pine planting in the area and designate some of Northland's remaining kauri forest as a national park was gathering strength. As New Zealanders seized the opportunity of the centennial to closely examine their history and identity, their understanding of the history and identity of their trees emerged as a very mixed bag.

Given Parry's passionate plea to the 1937 conference to preserve New Zealand's native bush, it is reasonable to assume that he might have been conscious of the ironies inherent in the 1940 celebrations. There were, however, other centennial opportunities to bury the axe and the fire-stick and nurture a tree sense in New Zealanders. Most notable were two of the commemorative publications — the highly illustrated

Making New Zealand: Pictorial Surveys of a Century and the Department of Internal Affairs' newsletter, *New Zealand Centennial News*.[10]

The two-volume, 30-number series *Making New Zealand* records the country's first 100 years thematically and was influenced by the contemporary magazine-style series *Building America*, which illustrated life in the United States.[11] The editorial team was headed by Eric Hall McCormick (editor), included John Pascoe (illustrations editor), David Oswald William Hall (associate editor) and Oliver Duff (advisory editor), and was advised in matters of typography by the historian John Cawte Beaglehole. Parry and Heenan — as minister in charge and undersecretary, respectively — supervised, as did the National Centennial Historical Committee under the chairmanship of Member of Parliament and trade unionist James Thorn.

Various issues of *Making New Zealand* were written by authors commissioned by Heenan — often his own associates — or by members of the editorial committee, including Heenan himself. The series was, according to the *New Zealand Listener* in March 1941, the first history of the country to place itself firmly in the land itself.[12] As such, it was obligated to address the issues that pertained to its indigenous forests. In another startling set of ironies, among issues devoted to manufacturing, railways, racing, refrigeration, and sea and air and communications, the series includes two issues of particular interest to this examination of values ascribed to trees.

Number nine in volume one, *The Forest*, written by Alexander Hare McLintock, bears a romantic cover photograph by James Walter Chapman-Taylor depicting sunrays slanting through pristine beech forest. The text, however, leaps pragmatically into the fray in the first section, 'The Cost of Settlement', with an admonitory account of the devastation of the forest by fire and axe. The price of progress, McLintock argues, 'was the destruction of our native trees, and although we can point with pride to a century of achievement which has transformed New Zealand from a forested wilderness into a rich farming country, we have also to remember that some of our forest destruction has been hasty and ill-considered'.[13]

Aotearoa might boast of its scenery, but much of it had been mutilated. The legacy of 100 years of settlement, McLintock concludes, was landslides, floods, erosion, silting, spread of scrub and weeds, and a serious shortage of timber. Photographs, artwork reproductions, maps and sketches of pristine native forest, milling activities, fire, erosion, washouts and government pine plantations complete the lesson.

To provide a suitable finish for the *Making*

New Zealand series, Heenan persuaded naturalist, author, farmer and birdwatcher Herbert Guthrie-Smith to write *The Changing Land* as a retrospective on the effects of non-Māori intervention on the land. Heenan was an admirer of Guthrie-Smith's book *Tutira: The Story of a New Zealand Sheep Station*, which was first published in 1921. Guthrie-Smith delivered his text punctually, but when McCormick came to appraise his draft, he considered it too brief to be an effective survey. In the meantime, Guthrie-Smith had died.

In order to expand the text to the requisite length, McCormick grafted on appropriate sections of Guthrie-Smith's last work, *Sorrows and Joys of a New Zealand Naturalist* (1936), which is a no-holds-barred assessment of the destruction of the country's environment. *Tutira* describes an ecological survey of Guthrie-Smith's life on the sheep station north of Napier where he lived and farmed from 1882 until 1940. He was a meticulous observer and recorder of the changes that had occurred over that time to both fauna and flora on his land. The final, rather disjointed essay under Guthrie-Smith's name is founded on this empirical research and begins with the hypothetical removal of all animals and people from twentieth-century Aotearoa — a theme also expounded in *Tutira*.

The land would, Guthrie-Smith speculated,

restore itself: native fern, tussock and scrub, woodland and forest would reappear where soil and climate had formerly suited each of them. The few exotic survivors would be confined to domains of slip edges and banks of river silt. 'Ancient New Zealand', he writes, 'provided an example of untutored Nature in her wisest mood. To this land of green leaves, running waters, and blue skies — one of the fairest lands in the world — European navigators came late in the eighteenth century. From that time all was changed.'[14]

Guthrie-Smith's essay sums up the overall message of the two overtly environmental pictorial surveys that trees are essential to New Zealand's economic, aesthetic and environmental prosperity. Further, it cements the contradiction between the nation's centennial celebration of progress and the dire consequences progress had wrought on the environment.

The second centennial publication important to note in the context of arboreal heritage is the *New Zealand Centennial News*, a 15-issue newsletter produced by the Department of Internal Affairs between August 1938 and February 1941. Parry's signature frequently underlines an introductory column that directly addresses the readers, keeping them up to date and urging them to participate in centennial activities. The newsletter provided an opportunity to press the cause for trees, and Parry reported that he encouraged every provincial centennial committee and the many interest groups with whom he came into contact to consider tree planting as a commemorative event. 'Can we conceive any Centennial memorial more enduring, more beautiful, or of greater appeal to this and succeeding generations than the tree?' he preached to the converted of the Wellington Beautifying Society in a quote in the newsletter's first issue.[15]

After all, Parry continued, trees as district memorials are far more appealing than 'a monument of stone or marble which may be of indifferent aesthetic merit'.[16] Long after the pageantry is over and the exhibition at Rongotai has closed, he instructs readers of the ninth issue, the commemorative trees will remain as living reminders of the 'completion of New Zealand's first century as a British country and the inauguration of another era of progress'.[17]

Parry's promotion of tree planting was multi-faceted. In the *Centennial News* in 1938 he encourages unofficial competitions in 'beautifying operations', suggesting that communities pit streets or towns against one another, and emphasising the advantages of 'the creation and increase of natural beauty spread throughout the Dominion'.[18] In the following year he assures state cooperation for 'active friends of trees' who were attempting to restore the indigenous forests as protection from floods, writing that 'those enthusiasts are striving to make New Zealanders properly tree-minded for their own welfare. The instinct of self-preservation must induce New Zealanders to form strong enduring friendships with forests.'[19]

In 1940, under the headline 'Trees to Save the Country', Parry metaphorically draws on the nation's preoccupation with the war in Europe, insisting that forests must always be the 'vanguard of defence against the destructive forces of erosion'.[20] Aotearoa responded to Parry's 1938 call for the 'planting of trees, more planting, and still more planting',[21] so much so that the *Centennial News* was prompted to offer advice on how and where to plant commemorative trees, to warn centennial committees to finalise their planting plans and order their trees early from nurseries or face shortages of available plants, and to stress the importance of caring for the trees after the centennial. At the end of the celebrations it was clear that the planting of trees and the establishment of parks and playgrounds had been the most popular forms of centennial memorials.[22]

In a parallel thrust for more trees in the country, Christchurch Teachers' Training College agriculture and biology lecturer Lance McCaskill devised another centennial scheme, also presented at the 1937 Dominion Conference on Bush Preservation and Amenity Tree Planting. That scheme encouraged schools, and by association families, to commemorate the year by planting native trees, shrubs, grasses and herbs, the progress of which was tracked in the *Centennial News*.

In 1939, McCaskill had undertaken a Carnegie Travelling Fellowship in the United States, studying rural education, nature protection and soil conservation. The work of the United States Soil Conservation Service and the planting by the Civilian Conservation Corps of nearly three billion trees following the Dust Bowl events of the 1930s were particularly influential on his work in combating erosion. He was passionate about New Zealand's indigenous flora and fauna, and, while teaching at the Christchurch college between 1933 and 1944, had established an educational garden of more than 400 native species.

McCaskill was forthright in describing the motivation behind his centennial enterprise, writing in the second issue of the *Centennial News* that the first hundred years of the development of the country had been a period of destruction of natural resources that had left New Zealanders with a 'peculiar psychological heritage'.[23] He berates New Zealanders for believing that nothing that was native to the country should be tolerated and preferring to be surrounded by only the trees, shrubs, flowers and birds of Europe and North America. 'So long as this psychological heritage held sway, so long was the development of real nationhood delayed,' he argues. McCaskill's scheme introduced children to the history of the native plants in their particular home regions and was designed to help them understand the challenges of erosion and flooding exacerbated by deforestation. It also encouraged them to respect their native trees.

The government's *Education Gazette* provided authoritative horticultural articles to assist teachers, and their students began collecting wild plants, cuttings and seeds appropriate to their geographical locations for propagation in school plots.[24] Some schools were experienced in horticulture, having customarily worked with the State Forest Service raising eucalypts, pines and cypresses for transplanting, but this scheme represented a philosophical shift.

The new focus on native plants, McCaskill points out when reporting on progress of the scheme in the tenth issue of the *Centennial News*, introduced a historical motive to horticulture in schools, or at least removed the emphasis on the 'grossly material aspect of arboriculture'.[25]

The final issue of the *Centennial News*, published

in February 1941, revealed that more than 300,000 native and many thousands of exotic trees, shrubs, grasses and herbs had been propagated by schools during the centennial year.[26] In the same year, Parry reported to Parliament that the success of the government's encouragement to local authorities, institutions and individuals throughout the country was evidenced by the fact that 'some 220,000 trees were planted for this purpose. In addition, large numbers of trees were planted by educational institutions, beautifying societies, and other public-spirited organizations and by many thousands of private citizens'.

Historian Kynan Gentry argues that, above all else, 'the Centennial was not a "national" celebration, but rather a large number of regional, provincial and local celebrations of national themes, punctuated by a handful of more broadly national events'.[27] Not everyone could visit the exhibition in Wellington and not everyone was motivated to attend the touring art exhibition or read the pictorial surveys. Tree planting, however, could be done at a district level with an abundance of local sentiment and little expense.

By the end of the official centennial period, Parry announced in a 1939 edition of the *Centennial News*, 'every adult New-Zealander should be able to say truly that "I have planted a tree," or "I have helped to plant a tree"'.[28]

Think Like a Mataī

—

Colin D.
Meurk

The mighty mataī begins life as a coppery mess of tangled yet flexible twigs, assuredly alive in their springy resilience. Through unhurried centuries thereafter, contemplating the world as it passes by, it eventually becomes a giant of the forests of Aotearoa New Zealand. Its status, however, is not assured. It finds itself in a contested environment, competing for our admiration and appreciation with trees from other countries, often another hemisphere. Nevertheless, the mataī endures. We might learn from it.

Throughout history, trees and forests have attracted respect, reverence, deference, ideology, fear and even loathing. They provide critical landscape functions, character, legibility, identity and metaphor, but they can also get in the way of imported aesthetics and human-centred land uses such as agriculture and urban growth.

This love–hate relationship with particular trees and forests is partly because of their bulk. They are the largest visible organisms on the planet and they compete for space. The conventional concept of 'tree' has some dimensional bounds — that is, various height and form criteria — but fundamentally 'tree' is a social construct: What is the biggest thing, and what is bigger than me?

What is the relevance of rākau to place-making in a distinctly Aotearoa place (nationhood) or in local spaces (neighbourhoods)? What is their symbolic meaning and value to us, and how important is it to reference the local species for visual, ecological and cultural authenticity? We may monetise the ecosystem services provided by trees in production forests or in their urban amenity-enhancing capacity. We may fondly imagine that our own trees will, in a right-plant-right-place kind of way, come out on top in any such comparison, but in general continent-derived trees will perform better than locals by these metrics.

The competitiveness of these introduced trees and their ability to regenerate have been honed by more extreme environments which have fire- and mammal-driven ecosystems and by a larger gene pool than those of our altogether more kind and stable oceanic archipelago, with its less urgent avian-predator pressure. The deeper psychosocial importance of indigenous trees, however, is in emotional attachment, the localised meanings we derive from them and their vital contribution to place-making, not to mention their fundamental, intrinsic value.

How do we put a value on trees that will stand the test of cultural shifts, utility, pest invasion and what American ecologist James Miller refers to as 'extinction of experience'?[1] Loss of visibility inexorably leads to loss of awareness or attachment, loss of protectiveness and, ultimately, through negative feedback, attrition of existence. An authentic culture, founded on a visible natural and human heritage, is currently vulnerable to globalisation, fashion and misinformation.

At this existential crossroads, we need clearly articulated awareness and pride in our unique biodiversity. We need innovative ways of ensuring *our* rākau are integrated across *our* urban and rural landscapes as dominant symbols of *our* unique place and biogeography. How may our tree policies better align with ancient wisdoms, modern conservation theory and human needs for genuine identity, as well as economic sustainability? Our noble mataī tree embodies this long, reflective view of the universe, essential for a future of well-being that is based on history of place.

Trees are biogeographically, ecologically, economically and culturally important. They embody ecosystem services that broadly encompass utility, sustainability, resilience, landscape integrity and legibility, natural character, symbolism and public health. So how do we define them? A tree often has local characteristics of structure, dimension or stature and form, but Western colonies have typically had Eurocentric attitudes and definitions of 'tree' imposed upon them. It is conventional, or perhaps fashionable, to think of a tree as having a clean trunk and often a spreading deciduous canopy or crown. Smaller versions are framed as 'lollipop' trees. Dictionary definitions emphasise the tall, perennial nature of a tree with a trunk that is vertical for some distance up to the first branches. Tree transcends fern, conifer, lily, cactus and broadleaved plants. Wikipedia pins tree stature to 10 metres.[2]

But that can't be right, because the tree line is one of the classic bioclimatic boundaries and human thresholds between forest and alpine/arctic tundra.[3] Montane trees don't abruptly stop when they are diminished to less than 10 metres well below 'tree line'. They become smaller in response to increasing cold until, at a mean midsummer temperature of about 10 degrees Celsius, they cannot produce enough wood and bulk to exist in the frigid air, frozen soil and short growing season. At this point tree height is typically 2.5 to 3 or 4 metres. We see this in southern beech trees that form our highest timberlines and in heath-like *Dracophyllum* grass trees at our southern latitudinal limit of dwarf forest on subantarctic Motu Ihupuku Campbell Island.

We do need to distinguish between a tree species (one that potentially can grow into a tree given adequate conditions, achieving its approximately

3-metre potential) and a timber crop (greater than 10 metres).[4] For example, a shrub of the low alpine zone such as snow tōtara never grows more than 2 metres tall, but a mountain beech tree may be reduced to this under duress of cold or infertility. In Europe and North America, these stunted forms are termed krummholz, meaning 'crooked wood'. Nature is always more variable and diverse than fits our desire for neat, rigid categories.

'Tree' is a social construct, an imposing biological form bigger and stronger than a human, one that is sometimes imbued with supernatural powers. We can, more or less, walk beneath trees. In urban and rural industrial-cultural landscapes, a premium is placed on utilitarian, fast-growing, productive, tall, clean trunks, or sometimes dense hedge forms, and often deciduousness. These characteristics, selected and bred for generations in northern Europe, are, admittedly, perfect for shelter belts, street trees and avenues. This concept of the ideal tree was imported to Aotearoa in the mid-1800s for reasons of nostalgia, familiarity and the rapid reforestation of the cut-and-burnt centuries-old native forests.

However, our trees don't always conform to these received conventions. They are frequently bushy and multi-leadered (for example, pōhutukawa, kāpuka and kōwhai), rather than achieving the classic lollipop shape or towering stature of a redwood, giant gum or linden tree. Exceptions are kauri, kahikatea, tōtara, rimu, miro, mataī, pukatea, pōkākā, southern beech, rewarewa, tawa, horoeka and a few others, which we might reasonably call our 'noble trees'. These distinctions between the ideal tree and the realities of our more idiosyncratic and 'messy' native tree forms have affected their acceptability in our most visible streets and parks, critically interfering with criteria of value and local place-making.

T he first trees were primitive fern relatives that existed in the coal-forming Carboniferous era more than 375 million years ago. About 245 million years ago, these tree ferns were muscled aside by cone-bearing conifers, commonly recognisable in Aotearoa today as imported needle pines, cypresses and redwoods. The Monterey pine became the foundation of our timber industry once the merchantable native forest had been decimated, while the windswept Monterey cypress is familiar along our exposed coasts or as clipped shelter belts. Other conifers are abundant in forestry and amenity plantings.

Aotearoa, part of the now largely submerged

continent of Zealandia, was at the opposite end of the planet to where the needle- and scale-leaved, dry-coned conifers were evolving in fire-prone continents on fresh soils. These disturbances and fertility favoured rapid reproduction, establishment and growth. We, in the great southern Gondwana supercontinent, had our own lineages of conifers, including berry-fruited podocarps such as tōtara, rimu and kahikatea (which in due course fed our bush birds), and kauri, related to the South American monkey puzzle tree.

Unlike their distant cousins, these developed on often stable, poorer soils formed in eroded sandstones, ash, schists and granites under more oceanic or subtropical leaching climates. Flowers of ancient angiosperms from about 140 million years ago and, later, southern beech, were blooming across Gondwana. These ecosystems were free of voracious mammals until after Zealandia rafted away around 85 million years ago.[5] Instead, we settled on tortoise-like growth and laid-back longevity.

Through the Palaeogene era, 66–23 million years ago, other broadleaved trees dispersed around the fragments of Gondwana via land bridges and island stepping stones, or by drifting in the wind or on the sea. These were the ancestors of gums, wattles, proteas, casuarinas, laurels, kāmahi, fuchsia and horopito, along with the southern tree ferns, conifers and beech. From about 10 million years ago, mountain building and then glaciation tipped the climate out of the comfort zone for southern subtropical and fire-adapted plants, which mainly survive now in Australia and Africa.

Through the Cretaceous period there was also an explosion in grass-like plants — bamboos, palms, grasses, sedges, rushes, irises and lilies, and southern or pan-Pacific restiads, cabbage trees and New Zealand flax. We now know that there was no continuous hereditary line of Gondwanan beech, podocarps and other flora on the Zealandia archipelago that waxed and waned through the long Cenozoic era, between splitting off from Gondwana and the onset of glaciation. But enough land remained above water to accommodate our unique tuatara, ancient frogs and some plants.[6] Beech, kauri and podocarps apparently came and went along connecting island chains.

It is hard to imagine a world devoid of land mammals, but this was the case in Zealandia until some bats flew here and, a mere millennium ago, dogs and rats were brought in with the first human voyagers and hunters from Polynesia. Fast forward to the early 1800s, and the floodgates opened with European colonisation, bringing with it thousands of the temperate world's most potent species.

Imports have trebled the number of naturalised tree species in this land. Mammalian grazers and predators speed up ecosystem processes such that only competitive, prolific, fast-growing and stoutly defensive plants and prey can survive. In this absence of game-changing mammals, everything operates at a bird-lizard-insect-snail pace.

And thus, our biota was ill-prepared for the onslaught of continentally honed introductions. The natural ebb and flow of evolution, climate and land formation have changed the living landscape through millions of years, but this latest human incursion threatened to transform and overthrow our unique natural history in a mere geological tick. Without urgent restorative action, our point of difference in a homogenised world remains at grave risk.

Aotearoa is far from being a homogeneous canvas when it comes to environment or history, each place being characterised by its own tree or trees. We have been called the 'land of little landscapes': a country that changes visually and vegetatively along steep topographic and climatic gradients over very short distances.[7] Due to a pronounced rain shadow rainfall, for example, ranges from a drenching 12,000 millimetres per year along the crest of the Southern Alps to a parched 350 millimetres just 100 kilometres away in Central Otago. We don't quite have true deserts, nor do we have frigid polar conditions, so almost 90 per cent of the landscape is potentially capable of supporting trees, by our definition.

Latitudinally, we range from subtropical Raoul Island in the Kermadecs to subantarctic Motu Ihupuku Campbell Island, and these warm to cool and saturated to dry climates naturally support diverse patterns of forest types.[8] The most distinct bioclimatic boundary is the altitudinal tree line. This defines the alpine/subalpine threshold, notwithstanding that in our oceanic climate some northern continental trees can surpass the limits of our own rather softer varieties. Southern beeches form our highest, temperature-controlled tree lines at 1500 metres in the north.

Tree daisies reach less than 800 metres in altitude on Rakiura Stewart Island, while grass trees are at their upper limit near sea level in the subantarctic. Beech is absent in the far south and in the central South Island 'beech gap'. This feature is a consequence of the clean-out by piedmont glaciers 100,000 years ago and the sluggish migratory capability of southern beeches to refill the gap. In their absence, various grass trees, tree daisies, mikimiki, weeping māpau and podocarp shrubs create a scrub line. Both types of forest line peter out to an approximate 2–3-metre-high hedge.

Lowland subtropical to warm temperate climax forests on old leached soils are potentially dominated by kauri (our most massive tree), podocarps (our tallest trees) and some broadleaved flowering trees, with pōhutukawa on northern coastal headlands and mangroves in saline mudflats. Kauri is probably still readjusting to the post-glacial world, and more southwards migration is likely under contemporary climate change.

South of the Waikato–Bay of Plenty kauri line, lowland tree dominance swings to podocarps: rimu, miro, rātā, kāmahi, tree ferns and epiphytes in the western and southern rainforests; kahikatea (our tallest native tree at 60 metres), mataī, pukatea, pōkākā, mānatu, myrtle and tī kōuka on floodplains; and tōtara, mataī, kāpuka, houhere, kānuka and kōwhai on eastern drylands. Where southern beech occurs, it tends to dominate the uplands and southern South Island.

Moisture in Aotearoa varies along a west–east axis. Prevailing water-laden westerlies push up against the axial mountains, releasing their water in torrents, then the dried-out, warming föhn winds sweep down the eastern plains. This creates a knife-sharp contrast between the rain and cloud forests of the west and the dry woodlands (now induced grasslands) of eastern basins and plains. Eastern floodplains have many rainforest species (because their roots are bathed in a permanent supply of ground water), but the low air humidity here precludes other delicate elements of rainforest — tree ferns, filmy ferns, bryophyte cushions and epiphytic ferns, lilies, orchids and saucer-like foliose lichens.

All of our natural forests are 95 per cent ever-green, and 70 per cent of native trees are bird-adapted berry and/or nectar producers,[9] which again sets them apart from deciduous northern continental temperate forests, which are characterised by forest-floor bulbs, spring flowers and dry seeds for rodents and finches. A sample of the commonest exotic trees here shows that 65 per cent have dry fruits,[10] and whereas among the top 50 more than 35 per cent have fleshy fruits, this proportion declines to 25 per cent across the larger list. The lack of natural fire and continentally arid conditions through the latter history of Aotearoa means we missed out on classic fire-adapted, highly flammable species such as needle pines, gum trees, wattles, gorse and broom.

Our evergreen forest plants are somewhat shade tolerant, affording these habitats some protection from invasion by most northern deciduous, shade-intol-erant forest inhabitants, which rely on a pre-leaf spring or canopy disturbance for regeneration. Accordingly, the most invasive weeds in Aotearoa forests are those that emulate our native bird-dispersed, shade-tolerant evergreen species — monkey apple, privets, cherry laurel, yew, holly, ivy and, more recently, fatsia.

Trees perform a keystone role in terrestrial ecosystems around the world. As the biggest natural objects in the landscape, they provide an immense resource of foliage, fruits, seeds and nectar, and harbour countless invertebrates and microbes. Thus, lizards and birds of all dietary persuasions find a larder and safe nesting in trees and forests. Tree root systems, as a rule of thumb, are as extensive as their canopies and tap deep into the ground for life-giving moisture and nutrients. It is important to remember the massive microbial and invertebrate biomass that coexists with the roots and contributes to the carbon and nitrogen economy. Together, these create a living fabric with the weathered bedrock, binding and protecting this soil from slipping and sliding away or being eroded by rain or river.

From a singular, functional perspective, this explains the popularity of fast-growing exotic trees like poplar, willow and alder for hill and riverbank stabilisation, and pines and gums for dry ground. The nearest equivalents we have in the native tree flora for wet ground are tī kōuka, mānuka, mānatu, kahikatea, pōkākā, pukatea, karamū, kōhūhū, maire tawake and mikimiki, along with reeds, sedge tussocks, toetoe and harakeke. For dry slopes and plains, there are tī kōuka, houhere, kānuka, kōhūhū and mikimiki.

Sadly, it is wishful thinking that indigenous species in their natural place will be superior to imported exotics just because they evolved here. Things are not that simple. Yes, when we bring social, cultural and intrinsic evolutionary values into the service frame, native species are highly cherished. But values are still more often assessed in terms of their harvestable monetary profits — wood or fruit production — and so ignore the more esoteric values and externalities such as natural character, identity, eco-tourist cachet and downstream biosecurity costs of exotics.

Nevertheless, we can embrace the 22 benefits of urban street trees in general, calculated by American urban designer Dan Burden in 2006, regardless of their origin.[11] In this age of climate warming, perhaps the most universally critical of these physical values are shading, reduction of heat-island effects and carbon storage. I can't do better in defining the potential role of trees in a post-peak oil world than to quote from an editorial published in *New Scientist* on 16 March 2019: 'Everything that oil can do, wood can too. Handled smartly, it can be used as a biofuel and as an alternative to petrochemicals. It can also be processed into high-performance materials that are already replacing steel and concrete in buildings, and may do the same in planes and cars . . . wood could solve some of our biggest problems, from carbon emissions to plastic pollution . . . radically rethinking how we obtain energy, move ourselves around, build cities and use land.'[12] In New Zealand, the rot-resistant heartwoods of tōtara and mataī do not need the toxic preservatives required by exotic pine timbers — another environmental plus.

Trees and wood might power a future global economy and workforce, albeit with more forest-bathing and less retail therapy. More trees in clusters, patches, sanctuaries and corridors create wildlife stepping stones, landscape integrity and connectivity via biodiversity ripple (halo) effects, but also a social and emotional engagement halo through interactions with foraging wildlife and discovery of landscape legibility.[13] Overall, trees are major contributors to urban ecosystem services and well-being.[14]

Trees are evocative. We gaze in awe at the most majestic so-called noble trees, and their regal presence shrinks our human ego. That is the power they exert at a personal level, but trees have also been employed as symbols of national, imperial or political power throughout history. Bundles/fasces of willow poles carried by Roman legions bridged rivers, facilitated Roman conquest and symbolised Roman, and later fascist, power; yew trees were fashioned into long bows in Middle Ages England; the 'heart of oak' built the flagships of the British empire. Closer to home, tangata whenua, and more recently European settlers, have long recognised the qualities of power, strength and beauty embodied in the sturdy tōtara.

Today, it sometimes seems that city architects and planners continue to hark back to Eurocentric regiments of tall, erect, fastigiate tree forms that perhaps project power, order, cleanliness and purity. Similarly, the received fashion for highly manicured, sprayed and sanitised gardens and lawns appears to fulfil a need to control the untidy, messy, almost threatening disorder of wild nature or some disorder or lack of control in our lives. It may be that the human predilection for clean and open parkland is based on memory of our evolutionary origins in the savannah, and the threat of hidden dangers in deep, dark, dense forest from wild animals and, latterly, antisocial humans. These fears have been spelled out in fairy tales and now in safer parks policies. Nevertheless, landscape ecologist Joan Nassauer takes another tack by defining a perfect nexus of messy ecosystems as the embodiment of complex and nurturing life, within tidy frames that are a subtle nod to control.[15]

The creeping displacement of one Indigenous heritage by another contributes to extinction of experience and was until recently reinforced in Aotearoa by generations of arborists, park managers and landscape architects trained in European traditions. Loss of visibility inexorably leads to loss of memory, identity, attachment and protectiveness, further reinforcing extinction. The Eurocentric definition of an acceptable tree for streets and parks subliminally discriminates against our more 'unruly' indigenous forms. Yet trees are vital to natural character as the dominant natural object of internalised and projected cultural landscapes.

If we express *our* nature by placing diminutive, ephemeral tussocks at the feet of imperial, exotic trees, as city designers frequently do, this is symbolically worse than nothing. Such green fluff creates a new, imposed historical narrative that will continually reinforce a national inferiority complex. For as long as gum trees dominate Christchurch's Anzac Drive and maples dominate the city's memorial avenues and rivers, it is no wonder that its citizens suffer from, and institutionally perpetuate, that extinction of experience.

Visibility of our cultural touchstones in each

day of our lives is essential or we will displace our deep, rich history with a superficial and confused dimension from somewhere else. This is about landscape legibility; being able to read and know all layers of our natural and cultural history as we step through our space, thereby contributing to place-making, identity, health, well-being and even marketability of a reconfigured soft, yet authentic eco-tourism.

Over the past 50 years there has been a search for more historical meaning and discovery of our roots, and hundreds of community groups throughout the country are now engaged in restorative native habitat planting in their spare time. In addition, central and local governments generally now support this movement, despite the inevitable push-back from those who haven't yet made the connections.

In 2020, the citizen science platform iNaturalist NZ — Mātaki Taiao conducted an online natural history recording event through the level 4 Covid-19 lockdown. This was run under the #StayiNatHome NZ project.[16] The intention was to encourage people to observe plants, animals and fungi in their homes, yards and gardens, and during neighbourhood bubble walks along streets and in local parks. The results are an unprecedented exploration and inventory of the urban environment.

During the month up to 27 April, 2600 mainly urban dwellers across the country made 40,000 observations of 4200 species. Native trees attracted particular attention (perhaps from a more ecologically literate subset of New Zealanders, who recorded 43 indigenous and 23 exotic trees in the top 200 plant observations), with 57 native species recorded as natural/established/regenerating in less kempt gardens and waste places, or deliberately planted.

The comprehensive study of trees carried out in Auckland by Mike Wilcox in 2012 adds a further 14 indigenous tree species observed in nature reserves.[17] From the same #StayiNatHome NZ project, we have, in addition, an inventory of the 119 commonly observed introduced tree species in order of observation frequency.[18] Wilcox lists another 28 common or characteristic exotic trees of Auckland, and recent studies reveal that 60–70 per cent of planted trees in Auckland urban gardens, streets and golf courses are exotic.[19]

These data present different taxonomic and community perspectives on the tree flora of Aotearoa. Reality is always a balance between competing data, ideas, values, purposes and needs. We are confronted by functionally and visually contrasting mixtures of trees, and plants in general, of diverse origins. These associations are sometimes benignly referred to as 'recombinant' or 'novel' ecosystems.[20] Many exotic trees were introduced for a variety of well-intentioned reasons, but not all are valuable for production, benign or neutral. Many are now regarded as already or potentially serious invasive weeds that will inexorably transform the character, ecology and landscape of Aotearoa, alienating us from our whenua if we don't actively assert the primacy of our own.

This is not just a matter of heading for the hills and controlling wilding pines, which could otherwise transform the high country into a replica of the Rockies or European Alps. We must also begin rolling back the sycamores and monkey apples from parks and gardens, and proactively recalibrate the ratio of native to exotic trees in our most visible, and therefore influential, cultural landscapes. Spatial patterns and processes that reflect these values — in patches and corridors, the facultative matrix where people live and the embracing halo effect — need to be built into our cultural landscapes.[21]

When I started writing this chapter there was no such thing as the Covid-19 pandemic. Everything has changed now and yet nothing has changed. We can be sure, and have long predicted, that this is not a once-in-a-lifetime event that disrupts us momentarily on our quest for ever bigger and better business-as-usual. It is a harbinger of multiple emergencies that are looming on the horizon.

We must learn from this experience to respect and be kind to one another, and to the other life forms with which we are privileged to share this remarkable planet and upon which we are dependent for food, pharmacopoeia, identity and wisdom. It is time to adopt the principles of strong sustainability[22] and recognise that the economy sits within a social sphere enveloped by the biosphere,[23] not the other way around.

Trees are part of the solution. They can continue to satisfy our material needs and, unlike many other resources we consume, they can provide one that is continually regenerating — one that is renewable and recyclable. Trees possess social, cultural and spiritual dimensions, and we need more of them; the aspirational target of the New Zealand government to plant a billion trees by 2028 is just a start.[24]

The best time to plant a tree was 30 years ago; the second-best time is now. However, let's not bury our being under a blanket of someone else's heritage. It is too easy to propagate a billion Monterey pines, poplars and willows to cover the scars of erosion and past neglect of the whenua. It takes reflection and effort to reclaim our heritage, the mana of the people, our birthright and our Aotearoa New Zealand identity. This reawakening must also apply to the economy and industry, rewarding and empowering a gentler humanity, and stepping back from the rampant, winner-takes-all materialism indulged in by the West since the Second World War.

Trees are keystones to the health and wealth of our planet. They translate scale between puny humans and gigantic skyscrapers, and they reconnect us with nature through foliage, flowers and feathered foragers. They inspire the slow movement.[25] They embody a calm, sustainable, steady state, alongside ecosystem regeneration and renewal, and patience. The fast, flashy exotics succumb to succession, wind or fire, whereas our native trees evolved in a relatively undisturbed, benign environment where there was a subdued imperative to race to the top. It is time to breathe.

There is no better metaphor of this principle of patience and humility than our sometime mighty matai. In Aotearoa, we must assume our role and responsibility as kaitiaki to both harness carbon and protect our country's unique richness for the generations ahead. The laudable plan to teach the country's history to every child must not forget its natural history, but instead feed an ecological literacy that reframes our values around a more sustainable, harmonious future. Perhaps we may even offer a modest example to the rest of the world.

There are many worthwhile actions that people can take in their own lives, including planting native trees, walking in forests, eating healthily and local, and learning about the dynamic natural world — both its potentials and limits. One very important action essential for the health of our society and our planet, but often overlooked, is to ensure that treeconomics is represented at decision-making tables. Many of the emergencies facing our world are ecological in nature, but are often blurred by sociopolitical-cultural denial, inertia, and a knowledge base that is too narrow and disconnected. One of the enemies of peace and welfare is complacency. We need holistic ecologists and mataī tree-ists on every governance body.

Nearly a quarter of a century ago, we stated our socially progressive intention for decision-making to reflect triple, then quadruple, bottom-line thinking. Many boards, executives and governing bodies have taken up the challenge and are more diverse than previously, especially in their acknowledgement of Te Tiriti o Waitangi the Treaty of Waitangi. However, the one specific quadruple bottom-line voice that is absent, with few exceptions, is that of ecological science — representing one of the most crucial areas of expertise and pillars of sustainability.

We need to acknowledge now, before it is too late, the strong physical, political and spiritual links between trees and place-making, well-being, strength and power, alongside humility, steadfastness, wisdom, patient long-termism, sustainability, and kindness to people and the planet. Like never before, it's time to think like a mataī.

E Tata
Tope
e Roa
Whakatipu

———

Huhana
Smith

E tata tope e roa whakatipu — a forest is easy
to destroy but it takes a long time to grow.[1]

This whakataukī encapsulates many a personal or
group realisation reached by contemporary kaitiaki and environmental activists and it perfectly captures
a project in which I am involved in Horowhenua. For
many years, hapū of customary landholdings have been
returning Indigenous forest cover to a much-revered
ancestral lowland coastal area that stretches from Waiwiri
Stream (flowing from Lake Waiwiri, formerly known as
Lake Papaitonga, a dune lake at Muhunoa near Levin)
and along the coastal hinterland to the Waitohu Stream,
just north of Ōtaki.

By protecting the wet margins of waterbodies
with Indigenous trees, shrubs, reeds and grasses, we have
also been hydrologically reconnecting these dynamic
coastal dune lakes, dune wetlands, streams and river
systems to the Kuku–Ōhau estuary. Currents of ground-
water flow beneath our wetlands and dune lakes in coastal
Horowhenua, and through recent research we are coming
to understand more fully the wai manawa whenua as
subsurface waterways or artesian geomorphologies.

The terrestrial ecosystems here bear the legacy
of clear-felling and sawmilling of Indigenous forests,

drainage of swamps and peatland, stream channelling, grazing on the dunes and sand flats, and removal of ferns and native grasses. This all occurred between 1840 and 1963, until only 1 per cent of the original native vegetation remained on lands between the mountains and the sea in our Kuku, Horowhenua, rohe.[2]

As the region was transformed for pastoral farming and dairying, once connected waterbodies were severed from one another by drainage schemes or by waterway engineering for flood protection. We understand from recent Treaty of Waitangi claims research work and hearings held for the Waitangi Tribunal Porirua ki Manawatū Inquiry that historic land tenure changes, particularly between 1840 and 1870, resulted in complicated internal struggles between iwi and hapū.

Political disturbances over the lands and waterways here were largely generated by Crown and court actions, especially during the tumultuous period from 1860 to 1870, the decade during which the Native Land Court was developed and became active. Ancestral reactions to these changes in land tenure and to the loss of ties to waterbodies and former forest cover continue to reverberate and to have an impact on the customary rights of current generations and our roles in better decision-making and beneficial actions for our remaining natural integrity.

The impacts of long-term damage to the remaining ecosystems that are still valued as taonga today are ongoing. In my many roles as an artist, an iwi and hapū member/researcher and kaitiaki, and as a trans- and interdisciplinary academic, I meet with many specialists to communicate complex environmental problems based on kōrero-a-iwi, local stories of place. These narratives, combined with action to revitalise the health of ecosystems, have become a critical means to achieve change. My main focus in research and practice, however, is to benefit the future generations of our coastal region, and to be a change agent for the planet.

I see my key role as an embedded artist-kaitiaki located in Kuku, Horowhenua, guided by the collective past, present and future wisdom and intelligence of our iwi, Ngāti Tukorehe. I am also a hapū member of Ngāti Te Rangitāwhia, Te Mateawa and Ngāti Kapumanawawhiti (who are also linked to Ōtaki), and I have ancestral ties to Ireland, Scotland and England.

Since 1996, our associated hapū have led many major revitalisation and regeneration projects for damaged natural ecosystems. Hands-on, action-oriented kaitiaki have conducted microbial source-tracking of *Escherichia coli* (*E. coli*) faecal contamination in polluted waterways; replanted many Indigenous trees, shrubs, grasses, sedges, flax and reeds in their rightful places; surveyed the health of shellfish on our coastline for unhealthy levels of contamination from cattle effluent; reconnected and revitalised dune lake systems; investigated projects for mitigating coastal erosion; and developed ideas towards a sustainable harakeke fabric industry. Each effort has been an act of consolidated healing of our environmental systems within our hapū landholdings, which underpin our identity as related peoples.

I recall the words of Hohepa Kereopa, an incredible kaumātua and tohunga who I met back in 2000 at a Mātauranga Kura Taiao wānanga held within the Kaimai Mamaku forests near Tauranga, who articulated the role of a kaitiaki so well: 'The job of the kaitiaki is to keep the things of Creation safe. The return from this is the relationship you get with the thing you are protecting and the knowledge and learning that comes from that. When the world was created, everything was given full wairua and mana, like the trees for example, so that everything is its own master. So, if people want to exercise kaitiaki, they will first need to understand the value of all things, and the wairua of all things . . . they will know the effects and consequences of doing things to trees, or whatever. For us this does not mean being in charge . . . you don't go and tell the pipi how to live, you allow it to have the opportunity to live the way it knows best, and that is what kaitiaki is . . . it is about knowing the place of the things in this world, including your place in this world. When you get to that point, you realise that the thinking of all things is the same.'³

I am a research leader or principle investigator within the **Kei Uta Collective**, an expanding and contracting trans- and interdisciplinary research group of mātauranga Māori, contemporary art, design, landscape architecture, climate change science, ecological economics, geomorphology, hydrology and social integration specialists. The collective's knowledge and skills are harnessed to catalyse and activate research spaces to make change on the ground.

I am also an active member of the contemporary art and design group known as **Te Waituhi ā Nuku: Drawing Ecologies**. I see myself as a gatherer of Indigenous and non-Indigenous visual and knowledge specialists from around the world, who come together respectfully on Māori landholdings. Working closely with our hapū, according to the tikanga or kawa, this creative group negotiates, as American art historian and cultural critic T. J. Demos argues, complex 'interrelated processes of . . . global-scale, world-historical, and politico-economic organization of modern capitalism stretched over centuries of enclosures, colonialisms, industrializations, and globalizations . . . where capitalism evolved within and against nature's web of life'.[4]

As a group of like-minded people, we are traversing these diverse meanings within the realities of today's environmental decline. In following Kereopa's position, our local hapū kaitiaki and researchers have engaged with the complexities that exist between contemporary visual culture and customary Māori environmental worldviews of place, culture and nature. We have had to reimagine how our hapū and accomplices might overcome these complexities for an ecological coastal dune region that was once highly revered, but whose 14 hectares of wetland had, by 1999, been reduced to a cow-pugged, effluent-polluted and nutrient-enriched puddle with struggling native riparian vegetation.

Dysfunction among our people led to our hapū kaitiaki attempting to overcome this malaise in order to heal ourselves and our relationships with one another and with our ancestral place, and to build our knowledge of the place via planting the many, many trees and shrubs that are now transforming our coastal landscape. We now know that in experiencing the revitalising consciousness of Tāne-mahuta within our reforestation projects, in feeling the return of mauri that accompanies the planting of riparian areas alongside streams and waterways, or in witnessing how our iwi, hapū and whānau can clearly see the results of our action-oriented projects when entering coastal areas reforested with native species, together we are re-establishing once longstanding relationships with the natural environment. To realise this action, we had to find considerable resources. From a grounded iwi- and hapū-led basis, with multi-institutional support over many years, we are now drawing on contemporary art-making methods via locally led wānanga with Te Waituhi ā Nuku.

Te Waituhi ā Nuku joined our Kei Uta Collective in late 2018. At our first wānanga in early February 2019, we integrated a series of ideas that might accelerate better actions for healing Aotearoa New Zealand's environment, particularly its fresh water. We focused on developing more projects and ideas that had brought benefits over time to the ancestral coastal cultural landscape of importance to hapū of Ngāti Tukorehe and Ngāti Wehiwehi. Te Waituhi ā Nuku met again in Kuku in early July 2019 to reflect on what we had first activated.

In February 2020, we reconvened again at Tukorehe Marae, Kuku, to refine our initial concepts, with more artists joining the group from other institutions to add fresh ideas and approaches.[5] Monique Jansen from Auckland University of Technology is planning a comprehensive long-term artistic biochar project for Kuku, in which we will grow a carbon crop of fast-growing trees that will be converted into charcoal and activated through compost heaps, worm farms or gabion baskets. This carbon will be returned to the land to filter nitrates from the paddock run-off, specifically for the health of the Kuku and Waikōkopu streams. She also proposes to create three-dimensional charcoal objects that will work in a gallery context and that can also be returned to the land, thereby closing the loop.

Lisa Munnelly of Massey University uses intricate drawing, three-dimensional modelling and projection mapping to visualise the complex interrelationships that exist between land, nature/trees, water and people. Emma Febvre-Richards, also from Massey University, and Dean Merlino, a Melbourne-based musician, are long-term colleagues within the Drawing Open International Research collaborative. Together, they explore the sensorial disruptions to trees and nature created by humans. They track sounds in forests and spaces, and create digital gestures in material and digital drawings that highlight contemporary living impacts on nature. This work is allied with Febvre-Richards' dementia research, which attempts to overcome human disconnection to the environment via sensorial and digital means. Through her dementia research project, she aims to increase cerebral well-being.

Finally, Frances Whitehead, from the University of Chicago's Institute of Art and ARTetal Studio, is a civic practice artist whose contemporary work is all about shaping future cities. She works closely with trees and nature in different urban environments. Her practice is deeply engrained in mapping, climate change, post-humanism, counter-extinction and culturally informed sustainability.

Te Waituhi ā Nuku continues to employ our local hīkoi, or 'walking and talking the land', research method, and in recent years I have undertaken many similar hīkoi around the world. Across Ireland (including Maynooth in County Kildare, Cork city and Galway city), from Germany to France and Spain, and from Mexico City to Puebla and Oaxaca city, I have walked with a range of art, design, landscape architecture and environmental specialists. These walks have also included cultural geographers, artist friends on the streets of their cities and others at major contemporary art events. In Mexico I have walked with museum professionals who are also engaged in opportunities that create bonds between circum-Pacific countries.

In turn, at my university, Massey, I have drawn Toi Rauwhārangi College of Creative Arts to the attention of key staff from our Whiti o Rehua School of Art (and related alumni), who are leading a Māori and Pacific approach to Indigenous-led research that originated in Winnipeg, Canada. This digital-focused research partnership, called The Space Between Us, is a circum-polar collaboration of multiple university research entities.[6] We converged for our first face-to-face discussions at an Indigenous workshop in Sydney at the launch of the Indigenous-led NIRIN, the 22nd Biennale of Sydney, in early March 2020, before the Covid-19 global pandemic forced us to return to Aotearoa.

We then entered a longer lockdown, which physically and emotionally grounded us. During my lockdown in Kuku, I read a *New Yorker* article in which Kate Brown described how she shared a garden plot with others to grow food during the New York lockdown.[7] She talked of Covid-19 as not just a public health issue but also an ecological one. She wrote eloquently: 'The interconnectedness of our biological lives, which has become even clearer in recent decades, is pushing us to reconsider our understanding of the natural world. It turns out that the familiar Linnaean taxonomy, with each species on its own distinct branch of the tree, is too unsubtle: lichens, for example, are made up of a fungus and an alga so tightly bound that the two species create a new organism that is difficult to classify. Biologists have begun questioning the idea that each tree is an "individual" — it might be more accurately understood as a node in a network of underworld exchanges between fungi, roots, bacteria, lichen, insects, and other plants. The network is so intricate that it's difficult to say where one organism ends and the other begins.

'Our picture of the human body is shifting, too. It seems less like a self-contained vessel, defined by one's genetic code and ruled by a brain, than like a microbial ecosystem that sweeps along in atmospheric currents, harvesting gases, bacteria, phages, fungal spores, and airborne toxins in its nets . . . Self-isolation is key if we are to stop the pandemic — and yet the need for isolation is, in itself, an acknowledgement of our deep integration with our surroundings.

'To fully respond to what's happened, we need to reflect on the worldwide ecological networks that bind all us together. Wesley [the author's neighbour] and I will resume our work of growing and harvesting when this pandemic ends. I hope that we'll be joined by others, all over the globe, who are eager to tend the community garden that is our world.'

We are entering new phases of climate change research and action, mindful of how the global lockdown has strengthened our collective resolve to exact transformative change. I regard the global hīkoi I shared with other creative knowledge specialists as acts that bind the threads of connection between like-minded and active people, as explained in the phrase 'tuia te muka tangata', or 'bind the thread of humanity'.[8] I draw upon all my experiences to inform my workplace, my artistic practice and my research life. Each encounter emphasises the extreme importance of mātauranga Māori, our Māori understanding of whakapapa or intricate genealogical systems of connection, as fulfilling ways of knowing.

Tikanga and kaupapa Māori are critically important in revitalising, leading and assisting beneficial change. This is not only so that our Māori governance bodies can forge real-life implementation strategies and plan actions for the vulnerabilities we face within Māori coastal landholdings, but also to engender new regenerative models for Te Taiao, the Earth, for agriculture and economics, and for culture and social cohesion that benefit local communities and neighbouring towns and cities within our wider regions.

This attitude was made clearer when English 'doughnut' or circular economic analyst Kate Raworth recently teamed with Māori Indigenous knowledge and science leader Dr Teina Boasa-Dean to Indigenise a regenerative reality for Aotearoa. Boasa-Dean alerts us to the fact that there is 'no substitute for helping people become more connected with their ecology and surroundings . . . we must make room for Indigenous knowledge, explaining that mātauranga Māori is a technical science and its concepts are planets away from western science'.[9]

Thankfully, there are other Māori knowledge leaders like Boasa-Dean. Fifth-generation Māori astronomer Professor Rangiānehu Mātāmua, for example, readily shares Indigenous knowledge that was handed down to him as a taonga tuku iho by his grandfather and other whānau astronomers before him. At the urging of his grandfather, Mātāmua decided to purposefully share this collective intergenerational knowledge to help revive ancient astronomical traditions in Aotearoa. As he states, 'For me, Matariki is part of the decolonising of our division of time. It's reclaiming our traditional, environmentally driven timekeeping systems that allow us to interact with our environment and acknowledge the changing of the year.'[10]

It's not only these Māori knowledge leaders who are arguing for change, but also our rangatahi who are insisting that more effective, culturally responsive actions must be implemented *now* to halt deepening environmental and social decline. All participants within the Kei Uta Collective operate within our mātauranga Māori- and kaupapa Māori-guided research platforms. We function within an environmental management paradigm, while also working readily with Western science. We are committed to reclaiming and reframing ancestral mātauranga as essential knowledge of place.

We map our kōrero tuku iho or Māori oral narratives and acknowledge whakapapa as our genealogical connection to all things, both material and spiritual. We make sure we don't make liberal use of mātauranga Māori in a manner that runs the risk of distorting both context and content.[11] Over many years of well-funded Vision Mātauranga research,[12] our marae-based wānanga have become key forums through which we share community-generated conversations about the environment, with co-developed actions involving all participants for holistic well-being. Te Waituhi ā Nuku is also respectful of Indigenous knowledge development and its innovations. Each participant acknowledges that our hapū and Māori researchers are the basis upon which we can strengthen creative, value-based innovative potential.

The Māori researchers on the team recognise how our tūpuna sustained themselves and their economies over centuries; today, we reframe this knowledge development in order to extend whanaungatanga and enhance kaitiakitanga in a contemporary context. We have been further assisted by the decolonisation of kaupapa Māori theoretical research over the past 25 years and by applying research at the interface.[13] We demonstrate that the healing of whenua, awa and our wai manawa whenua — land, waterways and the subsurface geomorphology — by planting Indigenous trees, shrubs, plants or native grasses can provide transformative changes that ultimately enhance the well-being and cultural survival of Māori communities.

Through our work, we note environmental decline, biodiversity loss, climate change and all the effects on the human condition from a Māori perspective. Synthesis plays an essential role in combining the activity of present projects, and the new knowledge that our interrelated and ongoing projects generate. Our environmental projects are active expressions of kawa, kaupapa and tikanga, and are exercises of tino rangatiratanga.

My paintings highlight multifaceted agency as acts of healing, presented as radiating or glowing trees. They are *my* metaphorical approach to articulating and communicating the problems of environmental, social and cultural landscape decline. I also highlight the nine planetary boundaries.[14] My works may appear as Western art-influenced oil paintings on linen, but they are culturally determined, based on the creative and artistic intelligence of Ngāti Tukorehe.

The small circular work *Ka Rere ngā Manu Rangatira* is a commemoration of the tragic mosque shootings in Christchurch in March 2019. A heavily pollarded tree from a Christchurch street glows against midnight blue; 51 tūī fly free from its branches. Humans have a damaging tendency to exert control, both over other people and over nature. This is evident, for example, in tree 'training' or, in its worst manifestation, as ecocide.

The constrained, pruned, potted and controlled trees in *Hīkoitia te Ao* were selected from the hīkoi in the countries in which I have walked. Each of these trees radiates the experiences and conversations I had while I was away from Aotearoa. This work was also informed by 21 years of decolonising methodological research established by Linda Tuhiwai Smith and Graham Hingangaroa Smith — hence the bold letters

KA RERE NGĀ MANU RANGATIRA
Oil on linen
300 mm diameter
Private collection, Wellington

**HĪKOITIA TE AO/WALK
THE WORLD, 2020**
Oil on linen
1500 mm diameter
Collection of the artist

borrowed from a Tate Modern artwork placed before a shadowy elm sourced from the streets of London.

In these two works, cobalt and phthalo hues of blue, Payne's grey, black and titanium white predominate. They highlight my concerns about the inadequate global political, social, economic and cultural agency to accelerate the change we need to make.

These days, I only manage to put brush to linen when I can. More often, I tend to 'create' in more active and on-site ways rather than via the dealer gallery exhibitions I used to present twice a year. My work is inspired not only by our long-term research findings, but also by the experiences of site-based exhibitions or events presented to local communities to help them understand how co-created strategies might integrate complex issues, render solutions more accessible, and heal the relationships between local environments and their communities.

The collaborative research projects I am involved with always aim to mitigate the impact humans have had on the environment. By incorporating art, curatorial and design practices into kaupapa Māori and environmental research projects, my co-creators, members of my hapū research group and I have been able to hone our methodological considerations in order to achieve ecological and cultural restoration goals in a whole-of-person, whole-of-system context.

All our iwi- and hapū-led research and action with collaborators has been about empowering and supporting kaitiaki to reclaim ancestral mātauranga and to exercise tino rangatiratanga, whereby *we* take control of the changes to be made as authorities. We revive this mātauranga by analysing well-being as the basis for guiding our activities. Ultimately, we are reinstating the mauri to localised Māori cultural economies.

I recall the whakataukī that introduced this chapter, and note the insights and experiences that have been 'garnered from the ground' and from all others who also are dealing directly with their fragmented ecosystems. We clearly understand that a 'forest' is hard to regrow, but sustainability and environmental and human resilience can be enhanced through planned kaitiakitanga, through shared visions of art and design systems, and through the co-construction and co-design of strategies for better implementation of projects that effect change. Kaitiakitanga and regenerative development are complementary and recognise both economic and cultural imperatives.[15] We can rally efforts to effect significant improvements in these most uncertain of times.

Our
Lost Trees

———

Mels
Barton

In 2018, a research team at University College London studied the value of urban trees for their potential to store carbon and mitigate climate change. They found that trees in areas such as Hampstead Heath store up to 178 tonnes of carbon per hectare, compared with the median value for tropical rainforests of 190 tonnes of carbon per hectare.[1] As research such as this becomes more widely known, local authorities in Aotearoa New Zealand will find that retaining existing trees in cities is increasingly important to the nationwide bid to store more carbon, and as a consequence trees will gain more respect. In the meantime, however, our legislation prevents councils from protecting trees effectively and, as a result, they are being cut down every day and our cities are steadily losing their large trees.

It's hard to understand this approach, and Aotearoa is definitely well behind best practice in retaining an urban forest cover. Many local authorities around the world have made use of the freely available i-Tree Eco software application to calculate the economic value of the benefits their trees provide. In part because i-Tree Eco was created in the United States, more data are available from there than from other parts of the world. Some local authorities in the United States have combined their calculations to provide a figure for the total economic benefit gained from their municipal trees and compared the results with the economic costs of maintaining those same trees.

Research demonstrates that in New York City, trees provide US$5.60 (NZ$8.42) in benefits for every dollar spent on tree planting and care.[2] Street trees in Washington, DC, are estimated to produce annual benefits — defined as land value, quality of life, public health, hazard mitigation and regulatory compliance — of US$10.7 million (NZ$16.1 million).[3]

Closer to home, in the City of Melbourne (which makes up approximately 0.4 per cent of metropolitan Melbourne's landmass), amenity valuations have estimated that the 70,000 trees in its streets and parks are worth approximately A$700 million (NZ$756 million).[4] Melbourne's i-Tree Eco 2012 assessment of the 982 trees in Royal Parade, Collins Street, Swanston Street, Lonsdale Street and Victoria Parade shows that they removed 0.5 metric tonnes of air pollution each year at a dollar benefit of A$3820 (NZ$4126), stored 838 metric tonnes of carbon at a value of A$19,100 (NZ$20,630), sequestered 24 metric tonnes of carbon each year at a value of A$548 (NZ$592), saved A$6370 (NZ$6880) in energy costs each year through shading buildings in summer and providing solar access in winter, avoided carbon emissions by reducing energy use by A$114 (NZ$123) per year, and had a structural value (replacement cost) of approximately A$10.4 million (NZ$11.23 million).

In the United Kingdom, according to environmental social enterprise Treeconomics, the services provided by urban trees in Greater London are estimated to be worth £133 million (NZ$261 million) per annum. The carbon storage capacity of urban trees alone is valued at £4.8 million (NZ$9.4 million) per annum in Greater London, or £17.80 (NZ$34.93) per tree.[5] No such research has been undertaken in Aotearoa, but it would most certainly clearly show the value of the ecological services provided by our urban trees.

We are taking some small steps towards recognising this. In Auckland, for example, the Auckland Council Urban Ngahere (Forest) Strategy,[6] launched in 2019, recognises the social, environmental, economic and cultural benefits of the urban ngahere and sets out a strategic approach to knowing, growing and protecting it. Its aim is to increase tree canopy cover to 30 per cent (from 18 per cent) across the city's urban area.

Policy documents like this, produced by local authorities, are a positive step towards driving decision-makers to consider retaining and planting trees at every opportunity in our cities — providing the policies are implemented effectively at every level of local government. However, with a lack of legislative protection for existing trees and the daily threats that trees — particularly mature trees — face, policies like this will always be playing catch-up. It will be impossible to achieve the increased canopy cover targets unless legislative changes are made soon to protect the mature trees that we still have.

It is urgent because we are losing mature urban trees at an increasingly rapid rate. There are no official records of trees removed from private land, but The Tree Council, an independent, voluntary, non-profit, incorporated charitable society that has been serving the Auckland community since 1986 in the protection of trees, calculates that hundreds of these specimens are eliminated from our largest cities every week. It also considers that this trend will continue until legislative changes are put in place to stop the chop. Furthermore, The Tree Council estimates that we have lost one-third of Auckland's tree cover since the Resource Management Act 1991 (RMA) reforms were implemented in 2015 and general tree protection was removed. Until this tree protection measure is reinstated, the destruction will continue.

The value of trees in our cities is generally not well understood and, consequently, threats to our urban trees are growing. Rising house and land prices in urban areas, and the need to increase urban density in our cities due to population growth, are putting increased pressure on our green spaces and private gardens. The removal of general tree protection from the RMA in 2012 has meant that the elimination of the majority of trees from private urban land is now unregulated and requires no external conversation or legal consent. As landowners, we can simply walk outside with chainsaws and cut down 500-year-old trees on our sections without asking anyone's permission. Large, mature trees are disappearing regularly from New Zealand's urban centres. These trees, which provide the most benefits and services, are also most at risk because they take up more space.

Instead of being viewed as assets that enhance the value of a property — which according to international research they most certainly are — trees are now seen as potentially reducing the value of a property for redevelopment. They are now 'in the way', despite the fact that having mature trees on a property can actually increase its value. One of the well-documented, but not often appreciated, benefits of trees is their positive effect on our mental health and well-being. Hospitals and sanitoriums were once surrounded by beautiful plantings of trees so that patients could view them from their beds and get out among them while convalescing in a restorative and restful environment. Recent research has shown that the mental health benefits of large trees on private land for people who can see them from other properties, even at some distance away, are large and significant.[7]

This places a responsibility on landowners to retain and protect those natural assets for the benefit of others in their communities, but it is a responsibility not recognised in our law or in our local government planning schemes. There is no incentive, financial or otherwise, for us to retain and look after trees on our properties. If the motivation for buying a property is capital gain via redevelopment, or even just selling it on for potential redevelopment, then the trees will be at risk of removal due to the misguided perception that they will negatively impact the property's value.

The loss will only accelerate as building intensification continues unless the law is changed to recognise the value of retaining trees and we become better educated. We need to act now before it is too late for our urban forests.

A key obstacle to protecting urban tree cover in Aotearoa lies in the way in which sustainable management is legally interpreted under the RMA. The

Act is 'effects based', which means that if applicants can demonstrate that the effects their proposals will generate on the environment are not problematic, then they can gain consent. This, in turn, has led to the use of the term 'less than minor' — in relation to the effects on the environment — becoming the critical point of applications. The Tree Council finds it astounding that so many large effects can now legally be seen as 'less than minor' in order to enable development.

The Ministry for the Environment's *Resource Management Act: Annual Survey of Local Authorities 1999/2000* shows that public participation in decision-making is effectively kept to a minimum and that 95 per cent of consents are passed without public notification.[8] This means that there is no scrutiny or input beyond the consenting authority and no ability for members of the public to challenge the merits of the decision legally. The survey reveals that fewer than 1 per cent of consents are declined. The RMA relies on the development of national policy statements (NPSs) to enable decisions to be made consistently across the country, but many of these have never been developed as they are often politically charged. As I write in 2020, almost 30 years after the Act was passed, public consultation is finally taking place on an NPS for indigenous biodiversity to enable the protection of native fauna, flora and their habitats.

The RMA has undergone many changes, the most significant for urban trees being that made by the National government in 2012. This change removed 'general tree protection' from the Act, a clause that previously provided protection for all trees over a certain size and for certain species across the country. Now, trees are protected in our cities only if they are specifically identified or 'scheduled' in a local authority plan, or if the urban environment allotment (UEA) on which they stand is identified in their local authority plan.

A UEA, as defined by the RMA, is an allotment that is no greater than 4000 square metres, is connected to a reticulated water supply system and a reticulated sewerage system, has a building used for industrial or commercial purposes or as a dwelling house on site, and is not a reserve or subject to a conservation management plan or conservation management strategy. This change to the RMA came into force in September 2015, and since then there has been virtually no protection for trees on private urban land in Aotearoa.

The change has effectively prevented local authorities from protecting trees on private land. On public land, however, general tree protection does remain. In response to the change, Auckland Council, for example, has identified a significant ecological area (SEA) layer to map zones of high-value ecosystem that have been identified in its unitary plan on both private and public land and, therefore, is retaining further general tree protection on these sites.

Prior to the 2012 RMA change, local authorities traditionally maintained schedules of special trees — usually large trees, or trees with historic or high amenity values — which were individually listed in planning documents and therefore had retained general protection. The Auckland Unitary Plan's Schedule 10 Notable Trees currently contains approximately 6000 trees,[9] but an audit of the schedule undertaken by Auckland Council in 2017–19 identified significant inaccuracies and errors. Adding trees to the Auckland schedule is extremely difficult, as the bar for qualification is set very high: a tree must be a near-perfect example of its species, and/or be rare and of significant age or be historically important, plus it must be seen by large numbers of people in order to establish sufficient amenity value.

Few trees qualify, and those that do must cross the additional hurdle of the landowner's agreement to the scheduling. The ability of a private individual to veto the listing of a public asset means that most trees nominated by members of the public for scheduling fail to be added to the list. In addition, trees that qualify receive no interim protection until a formal plan change process (requiring public notification) to alter the district plan or unitary plan is completed to enable them to be added to the schedule. This can take years, and in the meantime the trees can be lost. Removal of trees protected on public land, in SEAs and via scheduling requires resource consent. However, the majority of these applications are not publicly notified and, as noted, are routinely approved. Urban trees are not safe even when they are legally protected.

A significant body of work on the values and benefits of urban trees has been carried out internationally, with useful conclusions to consider for urban Aotearoa. Results of tree-related investigations suggest that the benefits of urban trees are many and varied. Researchers have demonstrated the ability of urban trees to control air and surface temperatures, improve air quality, manage stormwater, reduce flooding, improve water quality, and benefit both our physical and mental health. Research into the effects of trees in cities refers to beneficial outcomes for sound absorption, traffic management, crime reduction, household energy efficiency, and biodiversity conservation and enhancement.

Despite the popular perception that trees are in the way of property redevelopment, real estate values increase when healthy trees form streetscapes and inhabit front gardens. A 2017 study focused on three Sydney suburbs found that a 10 per cent increase in a street's tree canopy could expand individual property values by A$50,000 (NZ$54,000) on average;[10] a 2011 study in the United States found general increases in land values of up to 30 per cent in places associated with the presence of trees and vegetation.[11] Why, then, with this wealth of international study into arboreal advantages in cities, are our own urban trees in jeopardy?

Population growth and urban intensification pose a huge threat to our urban forests. Mature trees take up space, and space is worth money when a site is being redeveloped for more dense living. The lack of general tree protection in the RMA means that the majority of trees on private land currently have no protection whatsoever. It's a situation that enables developers to have carte blanche to remove all the trees on a site in order to redevelop it, and in Auckland, at least, this is exactly what is happening.

The Auckland Unitary Plan enables and encourages urban intensification, with the densest development focused on transport nodes and city or sub-city centres. As properties change hands, particularly in the 'leafy suburbs', the old villas with their large gardens full of mature trees are being removed and replaced with low-rise apartments and terraced housing. The trees are simply in the way and the chainsaws are running hot. Legacy plans, such as the Auckland City Council District Plan, described amenity value goals for their different residential zones, and allowed for negotiation of the height envelopes or site coverages of developments in recognition of the retention of significant tree amenity that supported zone amenity goals. This approach of development trade-offs has been lost with the introduction of the Auckland Unitary Plan.

The push over recent years towards 'maintenance-free' properties has been another barrier to the protection of urban trees. People are busy with work, and their evenings and weekends are often packed full of activities with family and friends. There is less time available for managing a garden and many don't want to be bothered with the maintenance required. Tasks such as sweeping up leaves and dealing with seeds blocking gutters and drains are now often seen as unnecessary burdens by many urban dwellers, who do not understand the values and benefits their trees provide. They consider their trees to be a nuisance.

The Tree Council frequently attends consent application hearings for the proposed removal of scheduled trees, only to find that the reasons given for removal are nuisance factors to do with maintenance rather than any fundamental problem with the tree or with public safety. Fortunately, the RMA is clear that these are not adequate reasons for removal of a scheduled tree. The role of educating everyone on the values and benefits of retaining urban trees has fallen increasingly onto the voluntary sector, and without the necessary funding, resources and spare time, it is impossible for community organisations such as The Tree Council and Forest & Bird to do this work effectively.

Changing environmental conditions are increasingly putting urban trees under stress. Many mature trees are exotics and therefore have not evolved to grow in the conditions in which they now find themselves. This can be an advantage for some cooler northern hemisphere species such as European ash and European beech, which relish the warmer conditions in Aotearoa and adapt well. For other species, however, the faster growth rate our climate promotes can eventually become a problem. A case in point is the Monterey pine, examples of which are becoming safety hazards in many suburbs, having grown too quickly and having now reached a point at which their structure cannot sustain their size.

Climate change introduces another complication to the life of both urban natives and exotics. Our repeated drier summers and winters are loading further stress on trees that are already dealing with urban environments that do not provide them with ideal growing conditions. Add to that the increasingly wet springs, with conditions encouraging a proliferation of pests and diseases, and it is easy to see how some tree species may reach a tipping point that threatens their survival. Diseases such as Dutch elm disease, kauri dieback and myrtle rust are proving to be existential threats to the species they affect, and kauri and Myrtaceae species have now been officially listed as threatened under the New Zealand Threat Classification System. Under increasingly changing climatic conditions, these diseases — along with others that will arrive in Aotearoa in the future — are proving devastating for both native and exotic species.

In cities of escalating urban density in Aotearoa, it is likely that our lost trees will never be replaced and any new plantings will never be permitted to grow to maturity, by which time they will compete for space with new buildings. For cities such as Auckland to grow and remain healthy and pleasant places in which to live and work, trees are needed, and these trees need protection if they are to survive. We must act now before it is too late for our urban forest and this taonga is lost forever.

No Place
for a Tree?

———

Susette
Goldsmith

Picture a typical city wharf: robust, flat, barren, grey, essentially concrete. A utilitarian site, once an instrument of shipping, a restless, risky platform of loading and unloading, comings and goings. Now a foot and cycle route — sea to city, city to sea. Exposed to sun and wind, salt laden. No room for nature: no place for a tree. How is it, then, that Wellington's Taranaki Street Wharf is home to a thriving grove of leafy karaka? Why are they there? Whose idea was it? What do they mean? First, we need some background.

The New Zealand Company's survey team, led by William Mein Smith, sailed into Te Whanganui-a-Tara Wellington Harbour (Port Nicholson) on the *Cuba* on 5 January 1840, followed closely by the first combined wave of 811 settlers on the *Aurora* on 22 January 1840, the *Oriental* on 31 January, the *Duke of Roxburgh* on 8 February, the *Bengal Merchant* on 20 February, and the *Adelaide* and *Glenbervie* on 7 March. At the time there were eight major pā situated around the harbour — Waiwhetū, Pito-one, Ngā Ūranga, Kaiwharawhara, Pipitea, Kumutoto, Tiakiwai and Te Aro.

Historian Angela Ballara notes that a migration, known as Te Heke Paukena and comprising Te Āti Awa from Waitara, a part of the Taranaki iwi and some Ngāti Ruanui people, had moved to Te Whanganui-a-Tara in the mid-1830s and settled with the permission of the Ngāti Mutunga chief Ngātata-i-te-rangi.[1] The iwi could claim legitimate occupation of the harbour as a result of Ngāti Mutunga's transfer of rights to Taranaki and Te Āti Awa in November 1835.[2] These were the people of Te Aro Pā. They are the people most closely connected to the karaka grove.

The new Pākehā settlement began in disarray. Despite early efforts at building temporary huts, the settlers were largely dependent on the superior skills and knowledge of Māori for the provision of shelter, food and useful information. Smith's first plan to survey the flat land of the flood prone Hutt Valley had to be abandoned and his team shifted their work to the opposite southern side. They measured pā and kāinga, and set their pegs in the Te Aro gardens and urupā as though Māori occupation didn't exist. Māori protested by removing the pegs.

The land parcels were completed and duly numbered, and the settlers made their selections in the order established by the New Zealand Company. Reserve land was chosen for Māori without their consultation, and Te Aro Pā — whose owners had never signed the company's deed of purchase — was designated as the centre of commerce for the new town. The land south of the pā was intensively cultivated with gardens that climbed well into the hills, and the waters of the harbour itself and the nearby streams that drained into it and into Waitangi Lagoon to the east were rich food sources for the Te Aro people. Smith's subdivision grid, however, spread over it all.

Following the signing of the Treaty of Waitangi at Wellington in April 1840, and subsequent enquiries into the validity of the New Zealand Company's purchase, some Te Aro Pā land was reserved for its people. It was, however, much coveted by the city's developers and had been singled out as the site of a wharf. Of course, ownership and occupancy of this land would mean the people of Te Aro Pā could benefit from the development of the city, but the Native Lands Act 1865 scotched that idea. Māori land title was individualised as a result and more land was free for purchase.

Te Aro Pā reserve was surveyed into 28 lots and land sales began in 1873. Plans for reclamation of land in the Te Aro area were already afoot, and in 1874 the Wellington City Council was granted 28 hectares of the Te Aro foreshore and harbour for reclamation. While it is not known exactly when Te Aro Pā moved completely out of Māori ownership, the Waitangi Tribunal Wai 145 report of 2003 notes that in 1850 there were 186 people living at Te Aro Pā and by 1881 the population had shrunk to only 28, which demonstrated a community in 'a state of terminal decline' and unlikely to have lasted very long into the 1890s.[3]

Wellington's settlers had been assured by the New Zealand Company that their part of the country would be the first to increase significantly in value. Key to the town's success, they were told, was Port Nicholson, which was the best harbour in New Zealand and guaranteed by its central position to be the focus of the country's coastal and maritime trade. Many of the European colonists were merchants and had arrived with goods to sell, barter or export. Australian wool ships returning from England were expected to discharge English goods before transporting wheat, wine, olive oil, fruit and vegetables to then drought-stricken Australia. Emigrant ships from England, it was proposed, would return with spars, turpentine, whale oil and raw or manufactured flax. Brisk trade was anticipated and appropriate facilities were required.

Squeezed as it was between hills and the shore, Wellington offered little in the way of available land for commercial expansion. The answer to this dearth of foreshore land for development lay in a series of reclamations, the first of which came courtesy of a severe earthquake on 23 January 1855. The earthquake raised the seabed of the harbour as much as 1.5 metres, making reclamation a much less costly prospect, and lifted Te Aro flat to become what is now the Basin Reserve, enabling the land to be drained for building development.

Between the early 1880s and 1925, six reclamations were carried out on Te Aro foreshore, which most recently had been occupied by boatbuilders, warehouses and small businesses. When the work was complete in 1886, close to 9 hectares had been reclaimed and almost all were put to use for commerce. The Wellington Harbour Board was instituted in 1880 and systematically developed the harbour into a commercial port. This included the requisite wharves, docks, sheds and other accoutrements of trade in response to the various developments in the shipping industry that continued to contribute to the city's prosperity. There had been early proposals to construct a wharf opposite the northern end of Taranaki Street, but it was not until 1906 that the 152-metre-long and 33-metre-wide Taranaki Street Wharf was completed and went into operation, primarily for the transport of coal and timber.

For 50 years, the wharf was a busy hub of port activities, but in the 1960s the face of the harbour began to alter. Writer David Johnson argues that the changes came from several directions.[4] Commercialisation arrived, and along with it came new shipping methods, which required fewer workers and less land. Increased recreation and tourism had drawn attention to the shoreline. The citizens of Wellington wanted their waterfront back.

This desire to reclaim the waterfront was part of a global trend. Air transport of both freight and passengers had developed, fishing methods had changed, ships had grown bigger and inner-city renewal was under way on areas of abandoned water's edge. Boston and Baltimore had led the way in America at the end of the 1950s, and London's Docklands became the first large-scale redevelopment project in Europe in the mid-1980s. Closer to home, the revitalisation of the Rocks and Darling Harbour, both in Sydney, took place in the 1970s and 1980s, respectively, and Melbourne's Docklands was redeveloped in the late 1990s.

The 1960s were the start of decades of draft waterfront schemes for Wellington, protest, consultation, revised plans and further consultation. The construction of the inner-city motorway in the 1960s had cut a swathe through a large part of the colonial cemetery and destroyed many of the nineteenth-century cottages that lay in its path. The condemning of historic buildings as earthquake risks and a building boom in the 1970s and 1980s led to further demolition of many of Wellington's historic buildings, before the 1987 stock market crash ground it to a halt. The wreckage prompted loud protest from new organisations formed to fight for threatened buildings and nurtured a reappraisal by many people of what was left of the city's historic fabric.

In mid-1996, a moratorium was called on further development on the waterfront and a committee was formed to consult with the community and come up with a report. After a series of public meetings and group discussions, the committee reported that Wellingtonians considered that heritage was a key component of the harbour's uniqueness and should be a major focus for any further plan. Restoration of buildings, objects, artefacts and monuments should be a guiding principle, and heritage must include both pre- and post-European history and acknowledge the importance of the waterfront to Māori.

A plan drawn up by a multi-disciplinary team of architects and landscape architects, along with urban designers, planners and advisers, was unveiled in October 1997. The public was consulted once more, and the plan — with some modifications — was adopted by the Wellington City Council the following year. The Taranaki Street Wharf area was selected as stage one, and in May 1998 Athfield Architects and landscape architect Megan Wraight were commissioned to prepare detailed drawings.

B ack to the karaka trees. Post-European history was already much in evidence in the Taranaki Street Wharf area, and many of the buildings, objects, artefacts and monuments qualified for heritage status of various degrees. The challenge for the design team was how best to represent *pre*-European history when no vestige of Te Aro Pā — the most local of the original pā — was in evidence at the time.

Furthermore, the very ground of Taranaki Street Wharf on which any representation might be established had, in fact, been reclaimed from the sea with the wharf's completion in 1906, and would have been underwater when the last residents had left the pā in the 1890s. The design team's solution was the establishment of a grove of karaka trees.

Why karaka? The species is classified by many domestic gardeners, agriculturists and scientists as a weed because of its habit of vigorously colonising cleared land, and is generally acknowledged as being dangerous because of its poisonous fruit. Its selection for amenity planting, therefore, might be regarded as somewhat audacious. For the earliest settlers of Te Whanganui-a-Tara, however, karaka was an extremely valuable resource. A medium-sized tree with large, thick, shiny leaves, the species is easily identifiable shimmering in native forest and, therefore, useful as a navigational marker.

Perhaps karaka's most striking characteristic, however, is its orange fruit, the flesh of which is edible to humans. The same is true of the kernel, which was an important portable winter source of carbohydrates and protein for Māori — but only after the toxins it contains were removed following extensive roasting and soaking. In addition, the leaves can be used in rongoā for sealing wounds and treating cuts and grazes. While the species is generally considered to be endemic to Aotearoa, Māori narratives recount that it was introduced from legendary Hawaiki.

Karaka is believed to have originated in the northern regions of the North Island and to have been distributed and domesticated by Māori travellers,[5] and descriptions of nineteenth-century Wellington note that large orchards were well established in the region at the time. In his examination of the recording of Māori occupation in the Pencarrow survey district, surveyor and archaeologist Bruce McFadgen quotes from an 1859 entry: 'In the Wainuiomata Valley prior to and during 1859, the standing bush consisted of Rimu, Titoki, Kahikatea, Rewa Rewa, Maire, Rata, Hinau, "abundant" Tawa, Totara, and "other pines of the finest description". At a distance of about two and a half miles up the Wainuiomata Valley, the bush gradually gave way to a large grove of Karakas, extending for about half a mile. These eventually gave way to "small fern and fine grass". The floor of the Orongorongo Valley was for the most part covered in shingle and scrub. There was however, a Karaka grove just over a mile from the coast alongside section 4.'[6]

In an edited version of his notes on the south-western Wellington coast, archaeologist Peter Beckett describes karaka growing 'at any suitable spots along the coast' in 1890, and records a grove of trees west of the stream at Pariwhero Red Rocks 'covering about a quarter of an acre' and another 'about half a mile along the ridge north of Red Rocks' where 'in a basin at the head of a gully falling westward to the Red Rocks stream, was an area of about fifteen acres of dense growth of trees. They were a mass of gold fruit — a sight never to be forgotten.'[7]

A grove of karaka on Wellington's waterfront, the design team concluded, was particularly apt. It would denote arrival and encampment; it would represent pre-European history.

The trees installed on Taranaki Street Wharf were grown from locally sourced material propagated by the Wellington City Council's nursery and were planted in 1999, when they were about 1.5 metres tall. Despite some early doubts from critics who said they wouldn't grow, the trees have prospered and the few losses have been replaced successfully. The overall effect of the non-regimented grove is of a 'natural' addition to its urban surrounds, but its design is decidedly 'cultural'. This is because any notions of 'arrival' and 'encampment' applied to a group of karaka as far south as Wellington are synonymous with intentional planting, regardless of the local origin of the plants. It is also because the trees are growing on reclaimed land and have had to be planted in a specially constructed, below-ground concrete tank to ensure that their soil does not become soaked in saltwater.

Above ground, the trees are contained in a raised rectangular bed composed of locally sourced aggregates, the surrounds of which double as seats, and security lights are embedded in the soil. At first, the karaka were confined in a metal cage designed to both protect them and provide them with shelter in the early stages of growth. Now, as they have grown, they are limbed up regularly to allow views under their canopies and to ensure the space is safe at night. Small birds add to the trees' living presence by settling noisily in the evenings in the warmth of the swaying canopy.

Clearly, these karaka have little in common with the entries in the various national and local tree heritage registers, which are, in general, large, old and/or rare trees. As a 22-year-old, the Taranaki Street Wharf grove is also unqualified for inclusion in the general list held by Heritage New Zealand Pouhere Taonga, the country's leading historic heritage agency, concerned with historic places and historic areas. Similarly, it is unlikely to be classified by the agency as Māori heritage ngā taonga tuku iho nō ngā tūpuna, treasures handed down by our ancestors, either as an example of 'physical/tangible heritage . . . land-based places created, formed or shaped by earlier inhabitants', 'natural heritage places . . . where no human activity is evident', or 'intangible heritage places . . . where no visible feature or evidence is present'.[8]

However, the karaka grove — despite, or because of, its spiritual connections to Te Aro Pā, its association with the international development of waterfronts and its role in the rehabilitation of its site — is decidedly part of the heritage of Taranaki Street Wharf. This is finely crafted heritage, subtle and rewarding for those who understand it. However, few of the many Wellingtonians who regularly walk or cycle past the trees to and from work every day, or the thousands of tourists who traverse Taranaki Street Wharf between the Civic Square and Te Papa each year, remark upon the presence of the trees or pause to ponder their provenance.

The grove has become part of the place's furniture, as ubiquitous as the seagulls. Within its larger, leafy context, all of which is almost completely dedicated to native plants and encompasses Waitangi Park (to the east), Frank Kitts Park (to the west) and neighbouring Bush City (part of the Museum of New Zealand Te Papa Tongarewa), the addition of a grove of native trees — particularly 'weed-like' karaka — was unlikely to excite much attention. The events leading up to the establishment of the grove, however, verify that it was not intended merely to 'soften' the Anthropocene landscape of Taranaki Street Wharf with some greenery and it was certainly not the result of a value-neutral decision. The landscape needed to grow more heritage.

Taranaki Street Wharf's karaka grove is not the only specially fabricated element that celebrates Māori history in this heritage quarter of the city. The bronze *Kupe Group* sculpture just north of the karaka grove is a replica of *The Coming of the Maori*, a plaster sculpture commissioned for, and first exhibited at, the 1940 New Zealand Centennial Exhibition at Rongotai, and was installed on its present site in 2000. The neighbouring Te Raukura building was opened in 2011 just west of the trees and is an initiative of Taranaki Whānui ki te Upoko o Te Ika[9] and Wellington City Council. Its design includes Te Wharewaka (housing two ceremonial waka taua), Te Wharekai (the very popular Karaka Café) and Te Whare Tapere (a conference and events venue).

In 1996, a heritage trail, Te Ara o Ngā Tūpuna The Path of Our Ancestors, was developed, and pā, cultivations, historic sites, buildings and cultural landscapes throughout Wellington are now marked by a series of symbolic carved pou whenua created for the purpose. Like the karaka grove, the pou acknowledge the relationship of tāngata and whenua. The trail was expanded in 2006 to include Waitangi Park and Toenga o Te Aro, some remains of Te Aro Pā discovered the year before when a building in Taranaki Street was demolished and developers began earthworks for a new apartment complex.

In the busy urban landscape of Wellington city these small signposts, the karaka on Taranaki Street Wharf, the *Kupe Group*, Te Raukura and Toenga o Te Aro signify Māori historic heritage and have become part of everyday experience. Many of the traditional Māori sites have been built over or destroyed and other than the fragments of Te Aro Pā structures there is little left to observe; a sense of spiritual presence lingers on or near the very place where real history was made.

The Taranaki Street Wharf grove is not marked on Te Ara o Ngā Tūpuna and, as reclaimed land, can lay no claim to authenticity as a historic place.

However, it is entitled to acknowledgement in that it contrives both a physical presence rich in symbolism and a spiritual presence of another kind — an Anthropocene *genius loci*. This grove of trees is more than just a feature of urban design or an addition to the city's indigenous arboreal inventory. Its location within the Taranaki Street Wharf site of memories, which is imbued with both Māori and non-Māori histories, demands a more thoughtful treatment of trees — as heritage.

In te ao Māori, nature, people and the land are inextricably intertwined through a shared whakapapa. The establishment of the karaka grove may have had its genesis in an official recommendation, but through contemporary design and implementation the grove aspires to make both tangible and intangible connections with the first peoples of Te Whanganui-a-Tara, who transported the food-bearing trees to the region from their natural northern home.

This flourishing karaka orchard in its Anthropocene habitat represents a living heritage that is steeped in history and symbolic significance, and tightly bound to the identity of Wellington.

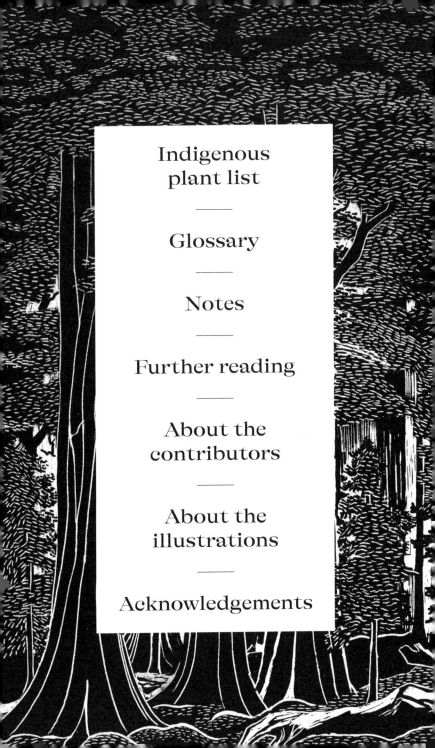

Indigenous
plant list

akeake *Dodonaea viscosa*

harakeke New Zealand flax, *Phormium tenax*

hidden spider orchid *Corybas cryptanthus*

hīnau *Elaeocarpus dentatus*

horoeka lancewood, *Pseudopanax crassifolius*

houhere lacebark, *Hoheria populnea*

horopito pepper tree, *Pseudowintera colorata*

hūperei black orchid, *Gastrodia cunninghamii*

hututawai hard beech, *Fuscospora truncata*

kahikatea white pine, *Dacrycarpus dacrydioides*

kāmahi *Weinmannia racemosa*

kanono *Coprosma grandifolia*

kānuka white tea-tree, *Kunzea ericoides*

kāpuka broadleaf, *Griselinia littoralis*

karaka *Corynocarpus laevigatus*

karamū *Coprosma lucida*; *C. robusta*

kareao supplejack, *Ripogonum scandens*

kauri *Agathis australis*

kawakawa pepper tree, *Piper excelsum*

kiekie *Freycinetia banksii*

kohekohe *Dysoxylum spectabile*

kōhūhū *Pittosporum tenuifolium*

kōkōmuka shore hebe, *Veronica elliptica*

kōtukutuku tree fuchsia, *Fuchsia excorticata*

kōwhai *Sophora microphylla*; *S. tetraptera*

māhoe whiteywood, *Melicytus ramiflorus*

maire tawake swamp maire, *Syzygium maire*

makomako wineberry, *Aristotelia serrata*

mamaku black tree fern/black ponga, *Cyathea medullaris*

mānatu ribbonwood, *Plagianthus regius*

manono *Coprosma autumnalis*

mānuka kahikātoa, tea-tree, *Leptospermum scoparium*

māpau māpou, *Myrsine australis*

mataī black pine, *Prumnopitys taxifolia*

mikimiki mingimingi, *Leucopogon fasciculatus*

miro brown pine, *Pectinopitys ferruginea*

New Zealand shiitake *Lentinula novae-zelandiae*

ngaio *Myoporum laetum*

nīkau *Rhopalostylis sapida*

patete seven-finger, *Schefflera digitata*

pirita green mistletoe, *Ileostylus micranthus*

pōhutukawa New Zealand Christmas tree,
    Metrosideros excelsa

pōkākā *Elaeocarpus hookerianus*

ponga silver fern, *Cyathea dealbata*

pukatea *Laurelia novae-zelandiae*

puku tawai *Laetiporus portentosus*

pūriri *Vitex lucens*

rangiora bushman's friend, *Brachyglottis repanda*

rārahu bracken, *Pteridium esculentum*

rātā northern rātā, *Metrosideros robusta*; southern rātā, *M. umbellata*

rewarewa New Zealand honeysuckle, *Knightia excelsa*

rimu red pine, *Dacrydium cupressinum*

taraire *Beilschmiedia tarairi*

tarata lemonwood, *Pittosporum eugenioides*

tawa *Beilschmiedia tawa*

tawai tawhai, silver beech, *Lophozonia menziesii*

tawhairaunui red beech, *Fuscospora fusca*

tawhai rauriki black beech, *Fuscospora solandri*; mountain beech, *F. solandri* var. *cliffortioides*

tāwiniwini bush snowberry, *Gaultheria antipoda*

tī kōuka cabbage tree, *Cordyline australis*

tī ngahere bush/forest cabbage tree, *Cordyline banksii*

tītoki New Zealand ash, *Alectryon excelsus*

toetoe *Austroderia* species

toru *Toronia toru*

tōtara *Podocarpus totara*; snow tōtara, *Podocarpus nivalis*

tutu *Coriaria arborea*

weeping māpou weeping matipo, *Myrsine divaricata*

whekī rough tree fern, *Dicksonia squarrosa*

Glossary

aruhe edible root or rhizome of the bracken fern

awa river

hāngī Māori method of cooking food using heated rocks buried in a pit oven

hapū descent group consisting of a number of extended families

hara transgression

hīkoi walk, procession, march

hoe canoe paddle

iwi Māori tribe, people, descent group; often refers to a large group who trace their descent from a common ancestor or migration canoe

kāinga village

kaitiaki guardian, steward, custodian

kaitiakitanga guardianship, stewardship, custodianship

kākāpō *Strigops habroptilus*, large flightless bird endemic to Aotearoa New Zealand

karakia prayer

kaumātua respected elders

kaupapa Māori Māori principles or philosophy

kawa protocols

kererū *Hemiphaga novaeseelandiae*, wood pigeon endemic to Aotearoa New Zealand

kete basket or kit, most often woven from flax

kiore Polynesian rat

kō long wooden tool used for digging

kōnini fruit of the native tree fuchsia kōtukutuku

kōrero-a-iwi local stories of place

kōrero tuku iho oral tradition, story or history

korimako *Anthornis melanura*, bellbird

kōtare *Todiramphus sanctus*, kingfisher endemic to Aotearoa New Zealand

mana term that denotes ancestral prestige authority, empowerment and influence. It is an intangible force that imbues the spiritual power and status of an individual attained through their genealogy

Matariki Pleiades, an open cluster of many stars in Te Kāhui o Matariki, with at least nine stars visible to the naked eye. The first appearance before sunrise of Matariki in the north-eastern sky, in the Tangaroa phase of the lunar month, indicates the beginning of the Māori year

mātauranga knowledge, wisdom, understanding and skill

mātauranga Māori knowledge, wisdom, understanding and skill; also relates to the passing on of traditional knowledge in a Māori cultural context

mauri life force

moa large extinct flightless birds of nine subspecies endemic to Aotearoa New Zealand

ngahere forest, bush

pā Māori fortified village, often situated on a prominent site such as a headland, hilltop or raised ground

Pākehā non-Māori person, New Zealander of European descent

Papatūānuku Earth, Earth mother

patu aruhe tool made of wood used to beat fern root prior to eating

pātua food baskets made of bark

pipi *Paphies australis*, a common edible bivalve with a smooth shell found at low tide just below the surface of sandy harbour flats

pīpīwharauroa *Chrysococcyx lucidus*, shining cuckoo, common in Aotearoa New Zealand

pou post, pole

pou whenua carved wooden post used to mark a place of significance

rākau tree, stick, wood

rangatahi youth, young people

Ranginui sky father

rohe territory

rongoā medicine

tangata whenua local people, Indigenous people of Aotearoa New Zealand

taonga tuku iho heirloom

taonga treasure

te ao Māori Māori world, Māori worldview

Te Ao Mārama the world of light, a crucial phase in the Māori creation narrative

Te Pō the darkness, a crucial phase in the Māori creation narrative

Te Taiao Earth, nature, natural world

te wao tapu a Tāne the sacred forest of Tāne

tauihu prow of a waka

tikanga cultural practices and ways of behaving

tino rangatiratanga Māori self-determination

tohunga cultural expert, priest or healer

tūī *Prosthemadera novaeseelandiae*, native bird endemic to Aotearoa New Zealand

tuia te muka tangata bind the thread of humanity

tuna eel

tūpuna ancestors

urupā burial ground

wai manawa whenua waterways

wairua spirit, soul

waka canoe

waka taua war canoe

wānanga intensive course of study or investigation, workshop, seminar

whakairo carving

whakapapa term for genealogy, the web of connections between people and the world around them, stretching back to the beginnings of the universe

whakataukī proverb or saying

whānau family

whenua land

whio *Hymenolaimus malacorhynchos*, blue duck, endemic to Aotearoa New Zealand

Notes

INTRODUCTION

1   William Cronon, 'Introduction: In Search of Nature', in *Uncommon Ground: Rethinking the Human Place in Nature*, ed. William Cronon (New York: W. W. Norton, 1996), 39.

A WALK IN THE BUSH

1   Elsdon Best, *Forest Lore of the Maori: With Methods of Snaring, Trapping, and Preserving Birds and Rats, Uses of Berries, Roots, Fern-Root, and Forest Products, with Mythological Notes on Karakia Used Etc.* (Wellington: Te Papa Press, 2005. First published 1942).

2   Murdoch Riley, *Maori Healing and Herbal: New Zealand Ethnobotanical Sourcebook* (Paraparaumu: Viking Sevenseas NZ, 1994).

3   Stanley Brooker, Conrad Cambie and Robert Cooper, *New Zealand Medicinal Plants* (Auckland: Heinemann, 1981).

4   Andrew Crowe, *A Field Guide to the Edible Plants of New Zealand* (Auckland: Penguin, 2004).

5   Robert Vennell, *The Meaning of Trees* (Auckland: HarperCollins, 2019).

6   https://www.landcareresearch.co.nz/tools-and-resources/databases/nga-tipu-whakaoranga-maori-plant-use-database.

7   https://www.nzpcn.org.nz.

8   Philip Simpson, *Down the Bay: A Natural and Cultural History of Abel Tasman National Park* (Nelson: Potton & Burton, 2018).

A LINE BETWEEN TWO TREES/
OBSERVATIONS FROM THE CRITICAL ZONE

1   David Haskell, *The Songs of Trees: Stories from Nature's Great Connectors* (Auckland: Penguin, 2018).

2   Pablo Neruda, *The Book of Questions*, trans. William O'Daly (Port Townsend: Copper Canyon Press, 2001).

3   Queensland Art Gallery/Gallery of Modern Art Facebook page, 'Anne Noble Discusses "Conversation: A Wonder Cabinet"', accessed 7 October 2020, https://www.facebook.com/QAGOMA/videos/anne-noble-discusses-conversatio-a-cabinet-of-wonder/2289640451265185/

4   Suzanne Simard, 'Net Transfer of Carbon Between Ectomycorrhizal Tree Species in the Field', *Nature* 388 (1997): 579–82.

5   Walter Benjamin, *Walter Benjamin: Selected Writings. Vol. 1: 1913–1926*, ed. Marcus Bullock and Michael W. Jennings (Cambridge, MA: Belknap Press, 1996), 62.

6   Ibid., 67.

AMONG TREES, AMONG KIN

1   Rudyard Kipling, 'The Song of the Cities', in *The Cambridge Edition of the Poems of Rudyard Kipling*, ed. Thomas Pinney (Cambridge: Cambridge University Press, 2013).

2   Aelian, *Historical Miscellany*, trans. N. G. Wilson (Cambridge, MA: Harvard University Press, 1997).

3   Mary Oliver, 'When I Am Among the Trees', in *Devotions: The Selected Poems of Mary Oliver* (New York: Penguin, 2017).

4   Peter Wohlleben, *The Secret Life of Trees: What They Tell, How They Communicate — Discoveries from a Secret World* (Vancouver: Greystone Books, 2016); Peter Tompkins and Christopher Bird, *The Secret Life of Plants: A Fascinating Account of the Physical, Emotional, and Spiritual Relations Between Plants and Man* (New York: Harper & Row, 2018); Robin Wall Kimmerer, *Braiding Sweetgrass: Indigenous Wisdom, Scientific Knowledge, and the Teachings of Plants* (Minneapolis: Milkweed Editions, 2013).

5   Kimmerer, *Braiding Sweetgrass*, 15.

6   Ibid.

7   Richard Powers, *The Overstory: A Novel* (New York: W. W. Norton & Company, 2018).

8   'Tales of Sweetgrass and Trees: Robin Wall Kimmerer and Richard Powers in Conversation with Terry Tempest Williams', 26 March 2019, Environment Forum, Mahindra Center, Harvard University, https://www.youtube.com/watch?v=9R4whGlL-EA

9   Ibid.

10   Rainer Maria Rilke, 'Ninth Elegy', *Duino Elegies*, in *The Selected Poetry of Rainer Maria Rilke*, ed. and trans. Stephen Mitchell (New York: Vintage, 1989).

THE GOLDEN BEARING

1   The *Aeneid* is a Latin epic poem in 12 books written by Virgil between 29 BCE and 19 BCE, and tells the legend of Aeneas, a Trojan hero who travelled to Italy.

2   This and following comments from Reuben Paterson are quoted from personal communications with the writer in 2020.

3   Pukekura Park in central New Plymouth is a botanical garden with a wide range of exotic and native trees. At Pukeiti, within native rainforest, is one of the world's largest and most diverse collections of rhododendrons.

4   Umberto Eco, *Faith in Fakes: Travels in Hyperreality* (London: Vintage, 1998), 43.

5   *The Golden Bearing* was accessioned into the Govett-Brewster Art Gallery's collection for five years and exhibited both outdoors and in the gallery during this time. The sculpture was exhibited on the Boatshed Lawn, Pukekura Park, 8 February–27 July 2014; the Rhododendron Dell, Pukekura Park, 14 December 2014–25 January 2015; the Govett-Brewster Art Gallery/Len Lye Centre 25 November 2017–28 January 2018; and the Founders Lawn, Pukeiti, 2 February–3 March 2018.

6   Eco, *Faith in Fakes*, 44.

7   Ibid., 7.

THE PECULIAR TREES OF AOTEAROA

1   'Index Seminum', Chelsea Physic Garden, accessed 9 October 2020, https://www.chelseaphysicgarden.co.uk/pages/category/index-seminum

2   Professor Stephen C. Sillett is the first Kenneth L. Fisher Chair of Redwood Forest Ecology at Humboldt State University, California; Marie E. Antoine is a botanist and lecturer at Humboldt State University, California; Robert Van Pelt is a forest ecology researcher and affiliate professor at the University of Washington, Seattle.

3   For example, see N. L. Stephenson, A. J. Das, R. Condit, S. E. Russo, P. J. Baker, N. G. Beckman, D. A. Coomes et al., 'Rate of Tree Carbon Accumulation Increases Continuously with Tree Size', *Nature* 507, no. 7490 (2014): 90–93; Becky Oskin, 'Old Trees Grow Faster Than Young Ones, New Study Shows', Huffpost, 16 January 2014, http://www.huffingtonpost.com/2014/01/16/big-trees-grow-faster-young_n_4609096.html; Peter Wohlleben, *The Hidden*

Life of Trees: What They Feel, How They Communicate: Discoveries from a Secret World, trans. Jane Billinghurst (Melbourne: Black Inc., 2015).

TREE SENSE OF PLACE

1   Jeffrey S. Smith, 'Introduction,' in *Explorations in Place Attachment*, ed. Jeffrey S. Smith (Abingdon: Routledge, 2018), 1.
2   Mindy Thompson Fullilove, *Root Shock: How Tearing up City Neighborhoods Hurts America and What We Can Do About It* (New York: New Village Press, 2016), 11.
3   Owain Jones and Paul J. Cloke, *Tree Cultures: The Place of Trees and Trees in Their Place* (Oxford: Berg, 2002), 87.
4   Quoted in ibid.
5   Quoted in 'For Dr Greg Moore, Defending Living Heritage Isn't About the Past, but Protecting Our Future', Foreground, 27 March 2020, https://www.foreground.com.au/environment/dr-greg-moore-significant-trees.
6   SmartView map available at https://smartview.ccc.govt.nz/map/layer/trees
7   Holly Best, 'Garden City', *The Occasional Journal*, March 2015, http://enjoy.org.nz/publishing/the-occasional-journal/the-dendromaniac/garden-city
8   For example, Phil Goff, then leader of the New Zealand Labour Party, said the city centre was 'like a war zone' ('Latest Updates: Christchurch Earthquake', *New Zealand Herald*, 23 February 2011, https://www.nzherald.co.nz/nz/ilatest-updatesi-christchurch-earthquake/A3GMU7KDEGXYCWS4RQBI2IIOGM; 'New Zealand earthquake aftermath: Your stories', BBC News, 25 February 2011, https://www.bbc.com/news/world-asia-pacific-12549427
9   Louise Beaumont, Dave Pearson Architects and Bridget Mosley, *A Conservation Plan for Hagley Park and the Christchurch Botanic Gardens. Volume One: History* (Christchurch: Christchurch City Council City Environment Group, 2013), 184.
10 Quoted in Liane Lefaivre and Alexander Tzonis, *The Emergence of Modern Architecture: A Documentary History, from 1000 to 1810* (Abingdon: Taylor and Francis, 2014), 334.
11 Quoted in John Knight, 'The Second Life of Trees: Family Forestry in Upland Japan', in *The Social Life of Trees: Anthropological Perspectives on Tree Symbolism*, ed. Laura M. Rival (Oxford: Berg, 1998), 203.

12   Andy Goldsworthy, press release for the *Garden of Stones*,
      2003, quoted in Jacky Bowring, '"To Make the Stone[s]
      Stony": Defamiliarization and Andy Goldsworthy's Garden
      of Stones', in *Contemporary Garden Aesthetics, Creations
      and Interpretations*, ed. Michael Conan (Washington, DC:
      Dumbarton Oaks Research Library and Collection, 2007), 181.

13   Will Harvie, 'Among the Ghost Houses: Walking
      Christchurch's Residential Red Zone', *The Press*,
      17 November 2018, https://www.stuff.co.nz/the-press/
      canterbury-top-stories/108410698/among-the-ghost-
      houses-walking-christchurchs-residential-red-zone

14   Annette Wilkes, 'Dallington Places', *Dallington /
      Christchurch: Our Home* (blog), accessed 9 October 2020,
      https://www.dallingtonlife.com/places-1

15   Ibid.

16   Michael Hayward, '"Quake Outcasts" Finally Paid for
      Uninsured Red Zoned Homes', *Stuff*, 18 February 2019,
      https://www.stuff.co.nz/national/politics/110542920/quake-
      outcasts-finally-paid-for-uninsured-red-zoned-homes

17   David Fisher, 'The Big Read: Last on the Land', *New Zealand
      Herald*, 25 February 2016, https://www.nzherald.co.nz/david-
      fisher/news/article.cfm?a_id=191&objectid=11595039

18   Simon Palenski, 'The Lines That Are Left', *Bulletin*, no. 184
      (2016), https://christchurchartgallery.org.nz/bulletin/184/
      the-lines-that-are-left

19   Kelli Truda Campbell, 'The Shaken Suburbs: The Changing
      Sense of Home and Creating a New Home after a Disaster'
      (MSc thesis, University of Canterbury, 2014), 81–82.

20   Ibid., 82.

21   Quoted in Michael Perlman, *The Power of Trees: The
      Reforesting of the Soul* (Dallas: Spring Publications, 1994), 26.

22   Matthew Adams, *Anthropocene Psychology: Being Human in
      a More-Than-Human World* (Abingdon: Routledge, 2020), 115.

23   Ibid., 116.

24   Keith G. Tidball, 'Seeing the Forest for the Trees: Hybridity
      and Social-Ecological Symbols, Rituals and Resilience
      in Postdisaster Contexts', *Ecology and Society* 19, no. 4
      (2014): 25.

25   Amber Silver and Jason Grek-Martin, '"Now We Understand
      What Community Really Means": Reconceptualizing the
      Role of Sense of Place in the Disaster Recovery Process',
      Journal of Environmental Psychology 42 (2015): 36.

26   Ibid., 34.

27 Ibid.

28 Keith G. Tidball, 'Seeing the Forest for the Trees: Hybridity and Social-Ecological Symbols, Rituals and Resilience in Postdisaster Contexts', *Ecology and Society* 19, no. 4 (2014): 25.

29 Jones and Cloke, *Tree Cultures*, 86.

30 Best, 'Garden City'.

31 Adams, *Anthropocene Psychology*, 117–18.

BURYING THE AXE AND THE FIRE-STICK

1 'Conserving Native Bush: Policy for Future, Important Conference', *Evening Post*, 2 April 1937. Where not otherwise attributed, quotes in this chapter from Parry and other delegates at the conference all derive from this source.

2 'Centenary Trees, Lord Galway's Suggestion', *Evening Post*, 4 August 1936.

3 Parry's remark is echoed by the statistic quoted almost 80 years later by Dr Nikolas Stihl, chairman of the global Stihl handheld power equipment company, that New Zealand 'racks up more chainsaw sales per capita — both professional and consumer — than anywhere else in the world'. Nick Grant, 'Stihl's Global Boss Calls "Bullshit" on Govt's Climate Change Response', *National Business Review*, 26 February 2016.

4 'National Reserves: Introduction of Exotics: Expert Expresses Fears', *New Zealand Herald*, 3 April 1928.

5 Jock Phillips, 'Afterword: Reading the 1940 Centennial', in *Creating a National Spirit: Celebrating New Zealand's Centennial*, ed. Bill Renwick (Wellington: Victoria University Press, 2004), 276. The official government view was that the centennial was 'the observance of the one-hundredth anniversary of organized settlement and government in New Zealand'. 'Summary of Centennial Organization', *New Zealand Centennial News*, no. 1 (1938), 2.

6 H. H. Bridgman (director), *One Hundred Crowded Years: New Zealand's Centennial* (Miramar: Government Film Studios, 1941), https://www.youtube.com/watch?v=lFYy-l6aI5A

7 M. J. O'Sullivan, *History of the Royal New Zealand Institute of Horticulture* (Wellington: Royal New Zealand Institute of Horticulture, 1952), 90.

8 Joe Heenan to Harry Allan, 30 July 1940, *Heritage New Zealand General-Trees* 22004-001, vol. 2.

9   Post and Telegraph Department Acting Inspector General,
        'The Centennial Postage-Stamp Issue: No. 2. The Fourpenny
        and the One Shilling Stamps', *New Zealand Centennial News*,
        no. 8 (29 April 1939): 9.

10   E. H. McCormick (ed.), *Making New Zealand*: *Pictorial
        Surveys of a Century* (Wellington: New Zealand Government,
        1939–40); *New Zealand Centennial News* (Wellington:
        Department of Internal Affairs, 1938–41).

11   *Building America: A Photographic Magazine of Modern
        Problems* took the form of 24 magazine-style numbers
        concerned with the political, social, economic and cultural
        life of America. It was published between 1935 and 1942.

12   *New Zealand Listener*, 14 March 1941, 4.

13   A. H. McLintock, 'The Forest', in *Making New Zealand:
        Pictorial Surveys of a Century*, ed. E. H. McCormick
        (Wellington: New Zealand Government, 1939), 2.

14   Herbert Guthrie-Smith, 'The Changing Land', in *Making New
        Zealand: Pictorial Surveys of a Century*, ed. E. H. McCormick
        (Wellington: New Zealand Government, 1940), 3.

15   W. E. Parry, 'Centennial Tree-Planting for Posterity: A
        National Appeal', *New Zealand Centennial News*, no. 1
        (15 August 1938): 4.

16   W. E. Parry, 'Desirable Memorials','' 3.

17   W. E. Parry, 'All Together for the Centennial', *New Zealand
        Centennial News*, no. 9 (29 May 1939): 1.

18   W. E. Parry, 'An Opportunity for All', *New Zealand Centennial
        News*, no. 2 (15 September 1938): 1.

19   W. E. Parry, 'Centennial Tree-Planting', *New Zealand
        Centennial News*, no. 8 (29 April 1939), 1.

20   W. E. Parry, 'Trees to Save the Country', *New Zealand
        Centennial News*, no. 14 (15 August 1940): 26.

21   Parry, 'Centennial Tree-Planting'.

22   The final list of memorials approved by the National
        Centennial Committee, as at 31 March 1941, was: 'Tree-
        planting, parks, play areas, 59; Plunket and rest rooms, 31;
        historical publications, 27; public halls, 17; swimming baths
        and pools, 14; Maori meeting houses, 14; memorial gates,
        10; beacons, cairns, obelisks, 5; motor camps, 4; community
        centres, 4; miscellaneous (libraries, rest-rooms, &c.,
        scholarships, tennis-courts, &c.) 10'. Of these, 95 had been
        completed and officially opened. *Appendix to the Journal of
        the House of Representatives*, 1941, H-22, 2.

23   L. W. McCaskill, 'Celebration of the Centennial in Schools: A

National Scheme for the Growth and Study of Native Plants', *New Zealand Centennial News*, no. 2 (15 September 1938), 8.

24   The *Aotearoa New Zealand Education Gazette Tukutuku Kōrero* was first published by the Government in 1921. Today, it continues to provide education professionals with education news and articles, and information on career development opportunities.

25   'Centennial Tree-Planting by Schools: A Widespread Programme of Progress', *New Zealand Centennial News*, no. 10 (27 July 1939), 20.

26   *Appendix to the Journal of the House of Representatives*, 1941, H-22, 2.

27   Kynan Harley Gentry, 'Associations Make Identities: The Origins and Evolution of Historic Preservation in New Zealand, 1870–1954' (PhD thesis, University of Melbourne, 2009), 264.

28   W. E. Parry, 'All Together for the Centennial', *New Zealand Centennial News*, no. 9 (29 May 1939): 1.

THINK LIKE A MATAĪ

1   J. R. Miller, 'Biodiversity Conservation and the Extinction of Experience', *Trends in Ecology and Evolution* 20, no. 8 (2005): 430–34.

2   'Tree', Wikipedia, last modified 6 October 2020, https://en.wikipedia.org/wiki/Tree

3   C. D. Meurk, 'Bioclimatic Zones for the Antipodes — and Beyond?', *New Zealand Journal of Ecology* 7 (1984): 175–81.

4   For many Aotearoa New Zealand tree metrics, see M. S. McGlone, S. J. Richardson and G. J. Jordan, 'Comparative Biogeography of New Zealand Trees: Species Richness, Height, Leaf Traits and Range Sizes', *New Zealand Journal of Ecology* 34 (2010): 137–51.

5   G. Stevens, M. McGlone and B. McCulloch, *Prehistoric New Zealand* (Auckland: Heinemann Reed, 1988).

6   D. E. Lee, W. G. Lee and N. Mortimer, 'Where and Why Have All the Flowers Gone? Depletion and Turnover in the New Zealand Cenozoic Angiosperm Flora in Relation to Palaeo-geography and Climate', *Australian Journal of Botany* 49, no. 3 (2001): 341–56; D. E. Lee, W. G. Lee, G. J. Jordon and V. D. Barreda, 'The Cenozoic History of New Zealand Temperate Rainforests: Comparisons with Southern Australia and South America', *New Zealand Journal of Botany* 54, no. 2 (2016):

100–27; M. S. McGlone, R. Buitenwerf and S. J. Richardson, 'The Formation of the Oceanic Temperate Forests of New Zealand', *New Zealand Journal of Botany* 54, no. 2 (2016): 128–55.

7   J. A. Hayward and K. F. O'Connor, 'Our Changing "Natural" Landscapes''', in *New Zealand Where are You? Proceedings of the 1981 Annual Conference of the New Zealand Institute of Landscape Architects* (Wellington: NZILA, 1981), 33–41.

8   C. D. Meurk, 'Evergreen Broadleaved Forests of New Zealand and their Bioclimatic Definition', in *Vegetation Science in Forestry*, eds. E. O. Box, R. K. Peet, T. Masuzawa, I. Yamada, K. Fujiwara and P. F. Maycock (Dordrecht: Kluwer Academic Publishers, 1995), 151–97.

9   C. J. Burrows, 'The Seeds Always Know Best', *New Zealand Journal of Botany* 32 (1994): 349–63.

10   See https://inaturalist.nz/observations?place_id=6803&project_id=stayinathome-nz&verifiable=any&iconic_taxa=Plantae

11   Dan Burden, *Urban Street Trees: 22 Benefits; Specific Applications* (Michigan: Gatting Jackson, 2006), https://www.michigan.gov/documents/dnr/22_benefits_208084_7.pdf

12   E. Wilson, 'Wood, Glorious Wood', *New Scientist* 241, no. 3221 (16 March 2019): 5.

13   Rotokare Scenic Trust Reserve, Rotokare Halo Project, accessed 22 October 2020, http://www.rotokare.org.nz/Projects/Rotokare-Halo-Project; C. D. Meurk and S. R. Swaffield, 'A Landscape Ecological Framework for Indigenous Regeneration in Rural New Zealand-Aotearoa', *Landscape and Urban Planning* 50 (August 2000): 129–44; C. D. Meurk and G. M. Hall, 'Options for Enhancing Forest Biodiversity Across New Zealand's Managed Landscapes Based on Ecosystem Modelling and Spatial Design', *New Zealand Journal of Ecology* 30, no. 1 (2006): 131–46.

14   C. D. Meurk, P. Blaschke and R. Simcock, 'Ecosystem Services in New Zealand Cities', in *Ecosystem Services in New Zealand — Conditions and Trends*, ed. J. R. Dymond (Lincoln: Manaaki Whenua Press, 2013), 254–73.

15   J. I. Nassauer, 'Messy Ecosystems, Orderly Frames', *Landscape Journal* 14, no. 2 (1995): 161–70.

16   See https://inaturalist.nz/projects/stayinathome-nz

17   Mike D. Wilcox, *Auckland's Remarkable Urban Forest* (Auckland: Auckland Botanical Society, 2012).

18   'Observations', iNaturalist NZ, accessed 9 October 2020, https://inaturalist.nz/observations?place_id=6803&project_id=stayinathome-nz&verifiable=any&iconic_taxa=Plantae

19   A. E. Chaffe, 'Do Urban Golf Courses Have the Potential to Contribute to the Sustainability of Urban Biodiversity in Auckland, New Zealand?' (MSc thesis, University of Auckland, 2016); S. Z. Huang, 'Auckland Street Tree Ecology: Current Status and Future Potential' (MSc thesis, University of Auckland, 2020); S. H. Peters, 'Composition and Structure of the Urban Forest, North Shore, Auckland' (MSc thesis, University of Auckland, 2012).

20   C. D. Meurk, 'Recombinant Ecology of Urban Areas — Characterisation, Context and Creativity', in *The Routledge Handbook of Urban Ecology*, eds. I. Douglas, D. Goode, M. Houck and R. Wang (London: Routledge, 2011), 198–220.

21   Meurk and Hall, 'Options for Enhancing Forest Biodiversity', 131–46.

22   Sustainable Aotearoa New Zealand Inc., *Strong Sustainability for New Zealand: Principles and Scenarios* (Wellington: Nakedize, 2009).

23   Kate Raworth, *Doughnut Economics: Seven Ways to Think Like a 21st Century Economist* (White River Junction: Chelsea Green Publishing, 2017).

24   Te Uru Rākau — Forestry New Zealand, *The One Billion Trees Programme: Our Future, Our Billion Trees* (Wellington: Te Uru Rākau — Forestry New Zealand, 2018), https://www.mpi.govt.nz/dmsdocument/31860/direct

25   Sten Nadolny, *The Discovery of Slowness*, trans. Ralph Freedman (New York: Viking, 1987).

E TATA TOPE E ROA WHAKATIPU

1   *Te Ao Hou*, no. 1, (Winter 1952): Inside cover.

2   John Rodford Wehipeihana, 'Sequent Economies in Kuku: A Study of a Rural Locality in New Zealand' (Master's thesis, Victoria University of Wellington, 1964), 27.

3   Paul Moon, *Tohunga Hohepa Kereopa* (Auckland: David Ling Publishing Ltd, 2003), 131.

4   T. J. Demos, *Against the Anthropocene: Visual Culture and Environment Today* (Berlin: Sternberg Press, 2017), 86.

5   For more information on Te Waituhi ā Nuku, including details of the participants and their practices, see https://www.drawingopen.com/te-waituhi-a-nuku-drawing-ecologies

6   The Space Between Us: Col(lab)orations within Indigenous, Circumpolar and Pacific Places through Digital Media and Design is a project that aims to enhance connections between Indigenous and Pacific peoples, settlers, and people of colour through public engagement of spaces via digital media, contemporary art and design means. These exchanges are collaborative, share intergenerational knowledge and build futures across oceans, particularly with Commonwealth nations/cities that share colonial histories. Key areas include Winnipeg, Toronto, and Iqaluit in Canada; Wellington and Auckland in Aotearoa; Gáivuona in Norway and Finland; Sydney and Melbourne in Australia; and cities of Hawai`i in the Pacific.

7   Kate Brown, 'The Pandemic is Not a Natural Disaster', *New Yorker*, 13 April 2020, https://www.newyorker.com/culture/annals-of-inquiry/the-pandemic-is-not-a-natural-disaster.

8   This is a line from the Toi Rauwhārangi College of Creative Arts karakia, 'Tuhia ki Runga'.

9   Dr Teina Boasa-Dean, indirectly quoted in https://www.projectmoonshot.city/post/an-indigenous-view-on-doughnut-economics-from-new-zealand

10   Dr Rangiānahy Mātāmua, quoted in Carmen Parahi, 'Why Matariki Matters', *The Press*, 20 July 2020, https://www.pressreader.com/new-zealand/the-press/20200720/281500753562311

11   See Penny Allan and Huhana Smith, 'Research at the Interface: Bi-cultural Studio in New Zealand, a Case Study', *MAI Journal* 2, no. 2 (2013): 133–49, http://www.journal.mai.ac.nz/content/research-interface-bi-cultural-studio-new-zealand-case-studyforpaper.

12   See Deep South National Science Challenge, 'Vision Mātauranga', accessed 13 October 2020 from https://www.deepsouthchallenge.co.nz/programmes/vision-matauranga

13   Allan and Smith, 'Research at the Interface'.

14   The nine planetary boundaries are stratospheric ozone depletion, loss of biosphere integrity, chemical pollution and the release of novel entities, climate change, ocean acidification, freshwater consumption and the global hydrological cycle, land system change, nitrogen and phosphorus flows to the biosphere and oceans, and atmospheric aerosol loading. See https://www.stockholmresilience.org

15   S. M. Smith, 'Hei Whenua Ora: Hapū and Iwi Approaches for Reinstating Valued Ecosystems Within Cultural Landscape', (PhD thesis, Massey University, 2007).

OUR LOST TREES

1   Phil Wilkes, Mathias Disney, Matheus Boni Vicari, Kim Calders and Andrew Burt, 'Estimating Urban Above Ground Biomass with Multi-scale LiDAR', *Carbon Balance and Management* 13, no. 10 (2018), https://cbmjournal.biomedcentral.com/articles/10.1186/s13021-018-0098-0

2   P. J. Peper, E. G. McPherson, J. R. Simpson, S. L. Gardner, K. E. Vargas and Q. Xiao, *New York City, New York Municipal Forest Resource Analysis* (USA: US Department of Agriculture Forest Service, Pacific Southwest Research Station, Center for Urban Forest Research, 2007).

3   Josh Foster, Ashley Lowe and Steve Winkelman, *The Value of Green Infrastructure for Urban Climate Adaptation* (Washington, DC: Center for Clean Air Policy, 2011), http://ccap.org/assets/The-Value-of-Green-Infrastructure-for-Urban-Climate-Adaptation_CCAP-Feb-2011.pdf

4   City of Melbourne, *Urban Forest Strategy: Making a Great City Greener: 2012–2032* (Melbourne: City of Melbourne, 2012), 22.

5   Wilkes et al., 'Estimating Urban Above Ground Biomass'.

6   Auckland Council, *Auckland's Urban Ngahere (Forest) Strategy Te Rautaki Ngahere ā-Tāone o Tāmaki Makaurau* (Auckland: Auckland Council, 2019), https://www.aucklandcouncil.govt.nz/plans-projects-policies-reports-bylaws/our-plans-strategies/topic-based-plans-strategies/environmental-plans-strategies/Documents/urban-ngahere-forest-strategy.pdf

7   Daniel T. C. Cox, Jonathan Bennie, Stefano Casalegno, Hannah L. Hudson, Karen Anderson and Kevin J. Gaston, 'Skewed Contributions of Individual Trees to Indirect Nature Experiences', *Landscape and Urban Planning*, 185 (2019): 28–34.

8   Ministry for the Environment, *Resource Management Act: Annual Survey of Local Authorities 1999/2000* (Wellington: Ministry for the Environment, 2001), https://www.mfe.govt.nz/publications/1999-2000.

9   https://unitaryplan.aucklandcouncil.govt.nz/Images/Auckland%20Unitary%20Plan%20Operative/Chapter%20L%20Schedules/Schedule%2010%20Notable%20Trees%20Schedule.pdf

10 Roger Swinbourne and James Rosenwax, 'Green Infrastructure: A Vital Step to Brilliant Australian Cities'

(AECOM, 2017), https://www.aecom.com/content/wp-content/uploads/2017/04/Green-Infrastructure-vital-step-brilliant-Australian-cities.pdf

11   Foster, Lowe and Winkelman, *The Value of Green Infrastructure for Urban Climate Adaptation*.

NO PLACE FOR A TREE?

1   Angela Ballara, 'Te Whanganui-a-Tara: Phases of Maori Occupation of Wellington Harbour c. 1800–1840', in *The Making of Wellington 1800–1914*, ed. David Hamer and Roberta Nicholls (Wellington: Victoria University Press, 1990), 25.

2   Ibid., 30. See also Morris Love, 'Te Ati Awa of Wellington', *Te Ara — the Encyclopedia of New Zealand*, 8 February 2005, updated 1 March 2017, http://www.teara.govt.nz/en/te-ati-awa-of-wellington

3   Waitangi Tribunal, *Te Whanganui a Tara Me Ona Takiwa — Report on the Wellington District. Wai* 145 (Wellington: Waitangi Tribunal, 2003), 342.

4   David Johnson, *Wellington Harbour* (Wellington: Wellington Maritime Museum Trust, 1996), ix.

5   Helen Leach and Chris Stowe, 'Oceanic Arboriculture at the Margins: The Case of the Karaka (*Corynocarpus laevigatus*) in Aotearoa', *Journal of the Polynesian Society* 114, no. 1 (2005): 7–28.

6   Bruce McFadgen, 'Maori Occupation of the Pencarrow Survey District as Recorded on Early Survey Records', *New Zealand Archaeological Association Newsletter* 6, no. 3 (September 1963): 120.

7   Peter Beckett, 'Some Notes on the Western Wellington Cook Strait Coast', *New Zealand Archaeological Association Newsletter* 6, no. 3 (September 1963): 139. Beckett's original notes were edited by John Daniels.

8   'Māori Heritage: Ngā Taonga Tuku Iho nō Ngā Tūpuna', Heritage New Zealand Pouhere Taonga, accessed 9 October 2020, http://www.heritage.org.nz/protecting-heritage/maori-heritage

9   Taranaki Whānui ki te Upoko o Te Ika is a collective group of iwi, including Te Āti Awa, Taranaki, Ngāti Ruanui, Ngāti Mutunga and other iwi from the Taranaki area.

Further reading

Adams, Matthew. *Anthropocene Psychology: Being Human in a More-Than-Human World*. Abingdon: Routledge, 2020.

Benjamin, Walter. *Walter Benjamin: Selected Writings. Vol. 1: 1913–1926*. Edited by Marcus Bullock and Michael W. Jennings. Cambridge, MA: Belknap Press, 1996.

Bensemann, Paul. *Fight for the Forests: The Pivotal Campaigns that Saved New Zealand's Native Forests*. Nelson: Potton & Burton, 2018.

Best, Elsdon. *Forest Lore of the Maori: With Methods of Snaring, Trapping, and Preserving Birds and Rats, Uses of Berries, Roots, Fern-Root, and Forest Products, with Mythological Notes on Karakia Used Etc.* Wellington: Te Papa Press, 2005. First published 1942.

Brooker, Stanley, Conrad Cambie and Robert Cooper. *New Zealand Medicinal Plants*. Auckland: Heinemann, 1981.

Cronon, William, ed. *Uncommon Ground: Rethinking the Human Place in Nature*. New York: W. W. Norton, 1996.

Crowe, Andrew. *A Field Guide to the Edible Plants of New Zealand*. Auckland: Penguin NZ, 2004.

Dawson, John and Rob Lucas. *New Zealand's Native Trees*, revised edition. Nelson: Potton & Burton, 2019.

Demos, T. J. *Against the Anthropocene: Visual Culture and Environment Today*. Berlin: Sternberg Press, 2017.

Gibbs, George. *Ghosts of Gondwana*, revised edition. Nelson: Potton & Burton, 2016.

Haskell, David. *The Songs of Trees: Stories from Nature's Great Connectors.* Auckland: Penguin, 2018.

Jones, Owain, and Paul J. Cloke. *Tree Cultures: The Place of Trees and Trees in Their Place*. Oxford: Berg, 2002.

Knight, Catherine. *Nature and Wellbeing in Aotearoa New Zealand: Exploring the Connection*. Ashhurst: Totara Press, 2020.

Park, Geoff. *Ngā Uruora: The Groves of Life — Ecology & History in a New Zealand Landscape*. Wellington: Victoria University Press, 1995.

Park, Geoff. *Theatre Country: Essays on Landscape and Whenua*. Wellington: Victoria University Press, 2006.

Perlman, Michael. *The Power of Trees: The Reforesting of the Soul*. Dallas: Spring Publications, 1994.

Power, Richard. *The Overstory*. London: William Heinemann, 2018.

Raworth, Kate. *Doughnut Economics: Seven Ways to Think Like a 21st Century Economist*. White River Junction: Chelsea Green Publishing, 2017.

Riley, Murdoch. *Maori Healing and Herbal*. Paraparaumu: Viking Sevenseas NZ, 1994.

Rival, Laura M., ed. *The Social Life of Trees: Anthropological Perspectives on Tree Symbolism*. Oxford: Berg, 1998.

Selby, Rachael, Pātaka Moore and Malcolm Mulholland, eds. *Māori and the Environment: Kaitiaki*. Wellington: Huia, 2010.

Simpson, Philip. *Dancing Leaves: The Story of New Zealand's Cabbage Tree, Ti Kouka*. Christchurch: Canterbury University Press, 2000.

Simpson, Philip. *Pōhutukawa and Rātā: New Zealand's Iron-hearted Trees*. Wellington: Te Papa Press, 2005.

Simpson, Philip. *Tōtara: A Natural and Cultural History*. Auckland: Auckland University Press, 2017.

Simpson, Philip. *Down the Bay: A Natural and Cultural History of Abel Tasman National Park*. Nelson: Potton & Burton, 2018.

Smith, Jeffrey S., ed. *Explorations in Place Attachment*. Abingdon: Routledge, 2018.

Stevens, G., M. McGlone and B. McCulloch. *Prehistoric New Zealand*. Auckland: Heinemann Reed, 1988.

Vennell, Robert. *The Meaning of Trees*. Auckland: HarperCollins, 2019.

Williams, P. M. E. *Te Rongoa Maori: Maori Medicine*. Auckland: Reed, 1996.

Wohlleben, Peter. *The Hidden Life of Trees: What They Feel, How They Communicate: Discoveries from a Secret World*, translated by Jane Billinghurst. Melbourne: Black Inc., 2015.

About the contributors

DR MELS BARTON is an environmental activist and advocate who immigrated to Aotearoa in 1999 after spending 10 years working for Environment Agency Wales. She now has her own environmental and communications consultancy, and has run the Kauri Rescue citizen science project, testing treatment methods for diseased kauri on private land, since 2016. She chairs both the Titirangi Residents and Ratepayers Association and the Waitakere Ranges Combined R&R Group, and is a member of the management committee for Revive Our Gulf. She is the secretary of Auckland's non-profit Tree Council.

DR JACKY BOWRING is Professor of Landscape Architecture at Lincoln University. Her main research interests are in the areas of landscape, memory, phenomenology and melancholy. She has explored these in both designed works and scholarly investigation, ranging from the memorial garden at Auckland's Holy Trinity Cathedral through to the books *A Field Guide to Melancholy*

(2008) and *Melancholy and the Landscape* (2016). The effects of Christchurch's earthquakes have been a significant focus for her recent design and research.

GLYN CHURCH is a plantsman and author of several books and many newspaper columns on the natural world. He studied at Pershore College of Horticulture in Worcestershire, England, and the Chelsea Physic Garden in London to gain a Master of Horticulture, before immigrating to Aotearoa in 1976. Since then he has worked in public parks, tutored for the Open Polytechnic, and owned and managed a 3-hectare display garden and specialist nursery in Taranaki, where he sells rare plants obtained from contacts around the world.

DR SUSETTE GOLDSMITH is of Ngāti Māhanga and Pākehā descent. She is an independent writer and editor of non-fiction, and Adjunct Research Fellow at the Stout Research Centre for New Zealand Studies, Victoria University of Wellington Te Herenga Waka. She has edited numerous museum, art gallery and scholarly books, essays and monographs, has published several natural and social histories, and has contributed articles and regular columns for diverse publications. Her scholarly research explores ways of perceiving, interpreting and safeguarding natural heritage in Aotearoa.

DR COLIN D. MEURK is an adjunct fellow at the University of Canterbury, adjunct senior lecturer at Lincoln University and research associate at Manaaki Whenua Landcare Research. His research interests are biogeography, ecological restoration and design, landscape dynamics, urban ecology and conservation biology. He has been a consultant ecologist for the post-earthquake Christchurch rebuilding and has major involvements with community restoration projects in and around cities, promoting integration of biodiversity within production landscapes. He has received various awards in recognition of his applied conservation work and projects, including the Christchurch perimeter walkway concept.

ANNE NOBLE is Distinguished Professor of Fine Arts at Whiti o Rehua School of Art at Massey University and one of New Zealand's most widely respected contemporary photographers. Her most recent work is concerned with human threats to natural biological systems, and has resulted in photography, video and installation projects involving close collaborations with scientists, musicians and curators. Her projects about bees featured in the 9th Asia Pacific Triennial of Contemporary Art in Australia in 2019, and recent exhibitions in New Zealand explore the language of the forest using infrared video and cameraless photography. She was made an Arts Foundation Te Tumu Toi laureate in 2009 and awarded the New Zealand Order of Merit in 2003.

MEREDITH ROBERTSHAWE is a public programmer and curator based in Te Whanganui-a-Tara Wellington. For the past 12 years, she has worked in public art, public programming and curatorial roles at Wellington City Council, City Gallery Wellington, Govett-Brewster Art Gallery/Len Lye Centre in New Plymouth and Aotea Utanganui The Museum of South Taranaki. She holds undergraduate degrees in geology and archaeology from Victoria University of Wellington and the University of Auckland, and a post-graduate diploma in museum studies from Massey University.

DR PHILIP SIMPSON is a botanist, ecologist and award-winning author of natural and cultural histories, the latest of which is *Down the Bay: A Natural and Cultural History of Abel Tasman National Park* (2018). He formerly worked with the Commission for the Environment and the Department of Conservation as a botanist. Philip was a founding member of Project Crimson, a community-based project established to arrest the decline of pōhutukawa, and in 2009 he was awarded the Loder Cup for his dedication to conserving and promoting

New Zealand's native plant life. He is a trustee of the philanthropic trust Project Janszoon, which is working to reverse the trend of ecological decline in Abel Tasman National Park and restore and preserve the park's rich wildlife.

DR HUHANA SMITH, Ngāti Tukorehe/Ngāti Raukawa, is Head of School at Whiti o Rehua School of Art at Massey University. She is a visual artist, curator and researcher who engages in major environmental, trans-disciplinary, kaupapa Māori and action-research projects. She is co-principal investigator for research that includes mātauranga Māori methods and sciences in order to actively address climate change concerns for coastal Māori lands in Horowhenua–Kāpiti.

ELIZABETH SMITHER has published 18 collections of poetry and was the New Zealand Poet Laureate from 2001 to 2003. She has published six novels and five short story collections, as well as journals and memoirs. In 2004, she was awarded an honorary doctorate of literature from the University of Auckland, and in 2008 she received the Prime Minister's Award for Literary Achievement in Poetry. Elizabeth's poetry collection *Night Horse* won the Ockham New Zealand Book Award for poetry in 2018.

KENNEDY WARNE co-founded *New Zealand Geographic* in 1988 and served as editor until 2004, when he decided to pursue his own writing and photography. He writes mainly for *New Zealand Geographic*, *National Geographic* and *E-Tangata*. He has written books on the world's disappearing mangrove forests (*Let Them Eat Shrimp*, 2011) and on the Tūhoe iwi (*Tūhoe: Portrait of a Nation*, 2013), as well as two children's books (*The Cuckoo and the Warbler*, 2016; *It's My Egg: And You Can't Have It*, 2017). He lives in Auckland, but prefers to spend his time in the overlooked and undiscovered parts of Aotearoa, especially its forests, mountains and multitude of islands.

About the
illustrations

NANCY M. ADAMS (1926–2007) joined the Dominion Museum (later the Museum of New Zealand Te Papa Tongarewa) in 1959 and became known as one of Aotearoa New Zealand's foremost botanists and botanical artists. A substantial collection of her work is in the Te Papa Archive.

Cover: *Forest Scene*, n.d. Museum of New Zealand Te Papa Tongarewa, CA000888/027/0002

Page 6: *Asteraceae — Olearia angustifolia*, 1972. Museum of New Zealand Te Papa Tongarewa, CA000888/003/0024

Page 16: *Myrtaceae — Metrosideros umbellate*, n.d. Museum of New Zealand Te Papa Tongarewa, CA000888/005/0027

Page 20: *Cupressaceae — Libocedrus bidwillii*, 1962. Museum of New Zealand Te Papa Tongarewa, CA000888/043/0021

Page 44: *Cordyline indivisa*, n.d. Museum of New Zealand Te Papa Tongarewa, CA000888/006/0088

Page 54: *Myrtaceae — Metrosideros robusta*, n.d. Museum of New Zealand Te Papa Tongarewa, CA000888/005/0039

Page 70: *Mount Richmond, Apiaceae — Aciphylla squarrosa*, n.d. Museum of New Zealand Te Papa Tongarewa, CA000888/002/0065

Page 84: *Unidentified tree species*, n.d. Museum of New Zealand Te Papa Tongarewa, CA000888/027/0001

Page 104: *Nothofagaceae — Nothofagus solandri* var. *cliffortioides*, n.d. Museum of New Zealand Te Papa Tongarewa, CA000888/008/0002

Page 124: *Nothofagaceae — Nothofagus*, n.d. Museum of New Zealand Te Papa Tongarewa, CA000888/012/0001

Page 142: *Cyathea dealbata*, 1963. Museum of New Zealand Te Papa Tongarewa, CA000889/001/0015

Page 168: *Nothofagaceae — Nothofagus*, n.d. Museum of New Zealand Te Papa Tongarewa, CA000888/012/0001

Page 190: *Cunoniaceae — Weinmannia racemosa*, n.d. Museum of New Zealand Te Papa Tongarewa, CA000888/002/0043

Page 206: *Nothofagaceae — Nothofagus solandri* var. *cliffortioides*, n.d. Museum of New Zealand Te Papa Tongarewa, CA000888/008/0001

Acknowledgements

The book now in your hands might never have come to fruition if it hadn't been for the foresight and encouragement of Massey University Press and, in particular, publisher Nicola Legat, who saw merit in the project from the first tentative proposal. I am deeply grateful to Nicola, to project editor Emily Goldthorpe, copyeditor Susi Bailey, designer Megan van Staden and all at Massey University Press who have enabled this team of tree advocates to have our say. The generosity of colleagues, friends and whānau who have shared their knowledge and photographs with us when needed is also much appreciated, as is the financial support of Boon and Techlam. Thank you all.

MASSEY UNIVERSITY PRESS

First published in 2021 by Massey University Press
Private Bag 102904, North Shore Mail Centre
Auckland 0745, New Zealand
www.masseypress.ac.nz

Page 19: Elizabeth Smither, 'Tree breath and human', originally
published in *The Lark Quartet* (Auckland: Auckland University
Press, 1999), courtesy of Auckland University Press.
Page 46: Pablo Neruda, extract from *The Book of
Questions*, trans. William O'Daly (Port Townsend:
Copper Canyon Press, 2001), courtesy of the Pablo
Neruda Foundation and Copper Canyon Press.

Design by Megan van Staden
Cover artwork: Nancy M. Adams, *Forest Scene*

A catalogue record for this book is available
from the National Library of New Zealand

Printed and bound in China by Everbest Printing Investment

ISBN: 978-0-9951407-4-5

The support of BOON and Techlam is gratefully
acknowledged by the publisher